A Century of American Steel

A Century of American Steel

The Strip Mill and the Transformation of an Industry

Kenneth Warren

LEXINGTON BOOKS
Lanham • Boulder • New York • London

Published by Lexington Books
An imprint of The Rowman & Littlefield Publishing Group, Inc.
4501 Forbes Boulevard, Suite 200, Lanham, Maryland 20706
www.rowman.com

6 Tinworth Street, London SE11 5AL, United Kingdom

British Library Cataloguing in Publication Information Available

Library of Congress Cataloging-in-Publication Data

Names: Warren, Kenneth, author.
Title: A century of American steel : the strip mill and the transformation of an industry / Kenneth Warren.
Description: Lanham : Lexington Books, 2019. | Includes bibliographical references and index. | Summary: "This study examines the development and implications of the wide continuous hot strip mill, a technological innovation that transformed the steel industry starting in the 1920s. It changed the profitability of companies, the number and skills of their employees, and the nature, scale, and location of their operations"— Provided by publisher.
Identifiers: LCCN 2019043072 (print) | LCCN 2019043073 (ebook) | ISBN 9781498576994 (cloth) | ISBN 9781498577007 (epub)
Subjects: LCSH: Steel industry and trade—United States—History. | Rolling-mills—United States—History. | Rolling-mill machinery—United States—History. | Organizational change—United States—History.
Classification: LCC HD9515 .W333 2019 (print) | LCC HD9515 (ebook) | DDC 338.4/76691420973—dc23
LC record available at https://lccn.loc.gov/2019043072
LC ebook record available at https://lccn.loc.gov/2019043073

Many years ago, the writer received encouragement and help from the late Joseph Malborn. This survey of the industry in which he worked is dedicated to his memory and to the ingenuity and application of others who, like him, labored to transform it.

Contents

Figures and Tables

FIGURES

TABLES

Preface

The two leading themes of this industrial study are the importance of continuity and the impact of major technological changes, which cause discontinuities. Today's highly automated, technically complex sheet and strip industry, which constitutes by far the largest sector of the steel industry, was developed in a preexisting pattern of works that had been formed by the small-scale sheet mill industry of the nineteenth and early twentieth centuries. That basis for today's strip mill industry included personal factors and company structures as well as geographical distribution. Time and again over the past hundred years technical advance and industrial innovation have caused a number of episodes during which these various inherited structures and patterns have changed. Four such periods of accelerated change stand out—the 1920s, when fundamental advances in technology were pioneered and applied; the 1930s, as prolonged depression forced much of the steel industry to adopt the new methods taken up by only a few in the previous decade; the 1960s, in which a number of new, more powerful and technically sophisticated strip mills were installed; and the years since 1990, in which—and again in particularly uncertain commercial circumstances—a new variant of the industry has developed.

Existing companies shared an exposure to a revolution in technology in the nation's most basic of industries. Some showed initiative in their reactions; others went under; in a few instances, opportunity was seized to break through and build a major new enterprise; in yet other cases, positive response awaited the pressure of necessity. Conditions favored the man, or team, ready to recognize and seize the possibilities of the times. For others, different considerations ruled out anywhere near the same level of entrepreneurship. I have tried to identify and follow through some of the circumstances that affected firms in such different ways.

I am grateful to AIST for three of the illustrations. The other photographs were given to me by Joseph Malborn. An earlier version of the text was kindly read by a former leader of the industry. He is in no way responsible for my interpretation.

Kenneth Warren, January 2018

Abbreviations

ACLC	Andrew Carnegie papers in Library of Congress
AISA	American Iron and Steel Association
AISI	American Iron and Steel Institute
AMM	*American Metal Market*
AS&T	American Sheet and Tin Plate Company
BAISA	*Bulletin of the American Iron and Steel Association*
BFSP	*Blast Furnace and Steel Plant*
IA	*Iron Age*
ICTR	*Iron and Coal Trades Review* (UK)
ISE	*Iron and Steel Engineer*
ITR	*Iron Trade Review*
JISI	*Journal of the Iron and Steel Institute* (UK)
TAIME	*Transactions of the American Institute of Mining Engineers*
TNEC	Temporary National Economic Committee
USS	United States Steel Corporation
USWA	United Steelworkers of America
YS&T	Youngstown Sheet and Tube Company

Introduction

Innovation and Continuous Rolling Mills

At various times in the modern history of the steel industry, it has been necessary to sharply increase the output of a particular finished product. The steel rail provided a good example in the later decades of the nineteenth century. Year after year track extensions were made and demand for rails increased, and in the United States on a greater scale than anywhere else. There were various responses—build more mills, redesign and replace existing mills, perhaps installing three-high rather than the older two-high mills, or maybe increase the throughput of an existing mill by providing it with more power. Some lines of advance reduced unit costs of production, but depended on large capital outlays. The degree of subsequent success depended on whether costs of production, market conditions, and the efforts of sales teams made it possible to operate the more efficient, highly productive but costlier plant at a high utilization rate. In the twentieth century, the sheet, tinplate and strip mill industries involved many of the same considerations.

Traditionally the price of thin flat-rolled products was higher than that of most other forms of ordinary, rolled steel. This reflected the labor-intensive nature of production, the average small size of plant and a general lack of fully integrated operations. Circumstances in the United States in the early 20th century pointed to a considerable long-term increase in demand for this class of product, but early attempts to mechanize sheet production proved commercial failures. Before the United States joined in the First World War, use of steel sheets in making automobile bodies was increasing rapidly, and it was gradually realized that it might be possible to use the technology of continuous rolling, already adopted in making bars, rods, and billets, in the production of sheet.[1] This might increase the quantity produced, at the same time improve its quality in meeting increasingly demanding specifications, and reduce unit costs of production. Initially those involved in the efforts to achieve these three aims were probably not fully aware of the conditions and consequences that would be associated with their achievement, but there was in fact a remorseless logic about the ways in which introduction of new processes affected the scale, structure, and location of the industry.

First, in dramatic contrast with the hand mills which until then had dominated this part of the steel industry, the new style of production would involve highly mechanized operations. Second, the high productivity of its processes, and the fact that they had become continuous rather than batch, meant output per plant was largely increased. As a result, in contrast with the old-style sheet industry, fewer centers of production would be required. Looked at another way, operators at only a small number of existing mill locations could justify the large investment needed to reequip for the new technology. The implications for the future of those that could not do so were not necessarily spelled out. Third, although for existing sheet and tinplate mills the relatively small tonnage and high price of their product meant that they did not need, and often could neither afford nor support, the installation of steel- and iron-making capacity, in order to be highly competitive in the new order supporting capacity in steel furnaces and blast furnaces was normally needed as a part of the same works. Unfortunately many existing sheet mills were located in places unsuitable for the large-scale, efficient assembly of minerals required in iron making.

Inevitably the new production processes were not adopted by all existing firms. Those which pioneered often prospered and grew into major concerns; companies which failed to respond so positively might find a refuge for a time in specialization or stagnated, withered, and died. Some major steel companies had been either not at all or only slightly involved in the production of thin flat-rolled products. In a number of instances, it was only after they had endured years of depression that they were persuaded to consider expanding into this more resilient sector. In a few cases, new companies were formed in order to enter the strip mill industry.

The availability of bulk supplies of sheet steel of a reliably superior quality—in terms of surface finish, uniformity of composition, and enhanced ductility—helped widen its use. It encouraged major changes in consuming industries, including a very rapid increase in the use of sheet steel in making automobile bodies, and leading on to the development of a new industry producing deeply pressed shapes. There was a large expansion in manufacture of steel furniture and kitchen goods.

A final consideration was—if the new process proved so superior, both technologically and commercially, would it not prove essential for all existing producers to switch to it if they wished to survive? If this happened, and particularly if, later, expansion of capacity in this sector proved greater than in various other traditional lines, would there not be a danger that the large-scale investment involved would result in the effects of an overexpansion of capacity? The implication of this, as time and again in the past with other finished products, would be underutilization of costly capital investment. In such circumstances, although customers might benefit from the modernization in steel, stockholders in steel companies would not gain a fair return for their outlays.

These few interrelated considerations underlay the establishment and expansion of what, from nonexistence when World War I ended, became within less than 20 years the most important part of the American steel industry. Previously, in many parts of the iron, steel and rolling mill industries, plant engineers had taken up innovations made in Europe, though they had often greatly improved on them. In dramatic contrast, the wide continuous hot strip mill was developed as a response to conditions in the home market, and was from first to last a product of American research and development and plant engineering. This remarkable achievement would eventually transform the industry worldwide. What follows traces its origins and course in the economy in which it was pioneered, mushroomed, and first grew to maturity.

In autumn of 1958, the eminent economic historian W. W. Rostow gave a series of lectures in the University of Cambridge on "The Process of Industrialization." Two years later, his ideas reached a wider audience through a book, *The Stages of Economic Growth: A Non-Communist Manifesto.* Rostow distinguished five distinct phases of national economic development—the traditional society, preconditions for takeoff, takeoff, a drive to maturity, and an age of mass consumption. There has been a great deal of debate about Rostow's "stages," but one aspect of his thesis is of key relevance to what follows. He dated the US takeoff into sustained economic growth to the time of the Civil War. Reconstruction was followed by the drive to maturity, and then, from the early 1920s, the American economy was the world's pioneer in entering the age of mass consumption. In each of the last two stages the steel industry played a key role, undergirding the whole process of national economic growth. However, the demands made on this "basic" industry differed profoundly in these two stages. As there was a considerable period of overlap and uncertainty about the time at which one stage passed over into the next, the signals sent to steelmakers by the needs of the wider economy seemed much less clear than they later were seen to have been.

The late nineteenth century and the years up to World War I were dominated by the requirements of capital-goods industries, those activities which provided the productive foundations for a rapidly growing and geographically expanding economy and society. Though important lines of production also served the needs of individual households, they were small in comparison with those supplying the necessities of infrastructure building and the producer sector generally, as in farming, mining, the building up and servicing of other manufacturing industries, and, most relevant to the steel industry, in providing for improvements in transport by land and water. Wire for bounding stock or arable land; pipe for the young oil and natural gas industries; plate for construction of barges, boats, and ships; beams for bridges, factories, offices and apartments; and rails for the apparently unending extension of the nation's network of tracks—these were then the dominant lines of steel consumption. In 1899, rails alone made up 28 percent of all rolled iron and steel production; together, structural shapes and wire rod added another 21 percent.

By the early twentieth century there were already signs of change. A leading signal was the slackening rate of construction of railroads. During the six years from 1901 to 1906, the average annual mileage added was 5,148, but over the next six years only 3,726 miles. In the same two periods annual production of passenger cars increased from 17,802 to 161,135. In the dawning new era, the requirements of the industries making consumer durables would dominate steel consumption. However, there was continuity as well as change. Rails were needed for track renewal; plate continued in demand for shipbuilding, tank construction, pressure vessels, and so on; and, as population and affluence increased, heavy structural shapes would be called for in apparently ever greater tonnages for building. Tube and pipe mills turned increasingly to supply the long-established but now rapidly expanding oil and natural gas industries. In addition to installing more mechanical aids to increase output in their labor-intensive sheet mills, some companies began to try replacing the normal two-high "pull-over" mill by a number of mills arranged in tandem, reducing the need for muscle power by mechanical transfer of steel from one stand to the next. In doing so they were trying to apply to sheet steel the continuous rolling principles already successful with wire rods or billets.

As the years passed, it became obvious that the largest and most sustained increase in demand was taking place in thin flat-rolled products. Tinplate served the canneries that processed truck crops, meat, fish, and then later the suppliers of drinks for often distant markets, in which these convenience goods were consumed by an increasingly affluent population. Passenger car and truck manufacturers and rapidly growing domestic appliance trades required ever-increasing tonnages of higher qualities of sheet and strip. By the 1920s, a new structure of production was becoming well established. Soon afterwards it became dominant.

Implications of these alterations in demand were seen in two types of technical change in the steel industry: small-scale but incremental alterations to existing plant and methods, and far-reaching replacement of one type of plant and process of production by something radically different. The increased call for thin flat-rolled steels was met by extensions to existing mills and large investment in new units of broadly similar type. Next some of these mills were mechanized. Then the processes of improvement were overtaken by far more sweeping innovations involving replacing "batch" production processes in both hand and mechanized mills by "continuous" rolling methods, centered in the revolutionary hot strip mill.

The problems which had to be solved in changing the technology in thin, flat-rolled steel were much greater than those in earlier successful movements to continuous rolling. In this field of plant engineering success with any product depends on careful coordination of roll speeds in successive stands. Imperfect coordination leads to a costly tendency either to tear or to buckle the product. Thin sheet steel is particularly subject to these hazards. Questions of roll contours, especially their concavity or convexity, also presented serious difficulties.

By the end of the 1920s, the wide continuous hot strip mill had been brought into successful operation. It proved no more than a first stage on a long journey, as progress continued within the new technology. As new, more efficient strip mills were constructed, those already in operation were rebuilt to give them greater capacity and improve the quality of their product. A few decades later another phase of development was marked by a so-called "second generation" of strip mills, equipped with much greater power and closer quality control. When growth in demand slackened, a number of older strip mills were closed. Then, during the last decade of the twentieth century, another distinct new stage was reached in which the output of the giant, fully integrated complexes in which most strip mills had operated since their introduction were supplemented by new mills associated with the much less fully integrated processes of the "mini-mills."

In addition to its effects on the economics of the business in which it is applied, any major technological change will transform supply chains, marketing patterns and company structures. It modifies perspectives on space and therefore changes the optimum locations for production. Incremental changes are by definition relatively cheap and preserve much of the scale, degree of integration and location of the existing plants. Major changes in technology cost much more and will be focused in fewer plants, affecting scale, integration and even location drastically. Both patterns of change have social consequences. Incremental technical advance may result in the retraining or part layoff of the workforce; more fundamental development may relocate production, and, though creating important new centers of employment, sweep away whole workforces, leaving old industrial communities stranded. In short, technical change, emerging from the laboratory or pilot plant, can wreak havoc on established patterns of production and the settled ways of human society. All this happened with the strip mill. Many thousands of skilled jobs were lost, but the much smaller number of unskilled or semi-skilled posts that were created were physically less demanding. Given the wide distribution of the old-style sheet industry in communities that were often small and highly dependent on a single firm, the transformation of the industry caused serious localized unemployment. In the context of the New Deal, some hard questions were raised concerning the human impact of technological change and its wider economic context.

It may be useful to summarize the steps that marked the progress to the first successful hot strip mill. They extended over more than a quarter century. The first failed attempt was European, in a mill in Bohemia in the 1890s. The formation of the United States Steel Corporation in 1901 created another opportunity for experiment, for the new group controlled over 60 percent of American sheet and tinplate capacity and could well spare one of its all-too-many mills for the search for improved methods. Between 1902 and 1910 pioneering efforts toward continuous sheet rolling were made near Pittsburgh and in the Shenango Valley. A final US Steel effort was made in 1917. In each of these instances, technical problems, extra ancillary costs, or increases in the proportion of inferior sheets more than canceled out other cost savings. A breakthrough finally came in the early and mid-1920s, as demand rocketed. In 1924, a first continuous sheet mill was brought into successful operation at Ashland, Kentucky. Three years later the same steel company bought a works at Butler in western Pennsylvania, which contained a mill rolling strip steel rather than sheet continuously. Ashland, and still more the Butler works, were the prototypes of the new strip mill; before the end of the thirties almost 30 continuous sheet/strip mills had been built in the United States. Each one had a much higher output than even the larger old-type mills; the new technology caused a concentration in the industry, in both plants and companies, and despite the increasing output of thin flat-rolled products, the workforce was severely reduced. By World War II, the scatter of the old-type mill industry, above all in small-town, interior Ohio, had been largely replaced by a new strip mill industry focused at existing major works near Pittsburgh, in the "Valleys," along Lake Erie, and in the Chicago district. A completely new center of production had been created in southern Michigan, and there were important outliers of it elsewhere.

After 1945 reconversion to peacetime production and massive expansion in demand was accompanied by the extension of the strip mill industry to the West, establishment of a second focus on the east coast, and above all by increased output from existing mills and the construction of new units west of the Appalachians. Then, in the next round of major innovation, which occurred between 1961 and 1970, eleven new so-called "second-generation" strip mills were built. They cost vastly more and together added 32.4 million tons of new capacity. As with the strip mills of the late 1920s and 1930s, there was a noticeable shift in location. Growth in demand for strip mill products reached a peak in the early 1970s. After that demand continued to increase, but less rapidly and certainly from year to year, accommodating both the new-generation mills and continuing output from existing works, most of which remained in being, and some of which were upgraded to meet the competition of the new mills. At the same time, imported sheet and strip products bulked larger in supplying American consumption. Output of flat-rolled products varied widely from year to year. In the 1980s, variations were even greater, demand generally was lower, and imports were higher. A few strip mills closed, but large-scale overcapacity remained. At the end of the decade, there occurred another, very different twist in strip mill technology. "Mini-mills"—steelworks without the backing of iron manufacture and all the mineral-production and assembly-processing operations which that involved, but instead melting either scrap or cold iron in electric arc furnaces, and further processing this metal not through primary rolling mills but by continuous casting—had already existed for decades. They had increased in importance, producing a wider range of finished products, and improving the quality of what they made. Now at last they began to produce strip mill products. Not constrained by bulk movements of raw materials, they were able to locate in places suitable for access to markets, and/or in rural or small town situations, where labor was abundant, cheaper, more tractable, and, after training, more productive than in long-established metal-making centers. In short there was in some respects a reversion to the locational considerations of the old-style sheet industry of the early years of the century.

Gradually improving the quality of its products through the 1990s and into the new millennium, the mini-mill industry rapidly increased its share of strip mill products. Reacting to this challenge, operators of second-generation mills and of surviving earlier strip mills spent liberally in trying to hold their own. As a result, competitive pressures increased, a number of mills failed and closed, and there have been major changes in corporate structure. In a world trading system in which there is much overcapacity, further closures of American strip mills seem inevitable.

Chapter One

Foundations

The Sheet and Tinplate Industries to the Late Nineteenth Century

In the last twenty years of the nineteenth century, the steel industry grew with exceptional speed, and from the early 1890s the United States was the world's largest producer. The 1899 output of all categories of rolled iron and steel was more than three and a half times that of 1879. The transition from iron to steel was almost completed, and the wrought iron industry collapsed. As a result, the 1899 tonnage of rolled steel was well in excess of ten times the 1879 figure. There were important changes too in the shares of the various products within this upward-spiraling total. Steel rails remained in the lead, but their share fell from 81 percent to about 28 percent; structural shapes were not made of steel in 1879, but 20 years later they accounted for about 10 percent of the rolled steel output. The main flat rolled products, plates and sheets—at this time unfortunately not separated in industry statistics—made up about 1 percent of the total tonnage at the earlier date, but 20 percent or so by the end. Of this last, hugely increased sector, plate was the more important member, supplying the huge demand for constructional material, railcars, shipbuilding and so on. Even so, output of sheet increased rapidly.

When J. P. Lesley published his remarkable survey of the American iron industry in 1859, he reckoned 60 percent of the rolled iron of all types was produced east of the Alleghenies. This same area made roughly 26,000 of the almost 37,000 tons of plate and sheet iron made in the United States.[1] In 1860, national output of sheet iron was about 11,200 tons (of which Pennsylvania made 7,000 tons) as compared with 20 times that tonnage for bars and 22 times as much for railroad iron. Ten years after the Civil War, a new list of rolling mills made clear that there had been both substantial growth and a pronounced westward shift in the iron and steel trade, including sheet iron. Forty-seven works were now listed as producing sheet. Of these, 14 were located in the East, and the other 33 were either west of the Alleghenies or in three instances in central Pennsylvania. West Virginia, Kentucky, Missouri and Michigan each had one sheet-making plant. In Pittsburgh, 13 works had sheet as one of their finished products, and this area was far and away the national focus of the sheet mills. There were six works in other parts of western Pennsylvania and seven in Ohio. The Shenango Valley in northwestern Pennsylvania and the Mahoning Valley in north eastern Ohio—the areas in the future often linked as the "Valleys" district—each contained two sheet works. Portsmouth, a later Ohio Valley center, already had two sheet mills, but Wheeling, where the nail business was dominant, as yet contained only a single large plant which turned out sheet as well as a number of other finished products.[2] Gradually, over the last quarter of the century sheet iron was replaced by sheet steel, often made by firms which had previously rolled iron, but increasingly by newly founded sheet steel companies.

The sheet iron industry had grown from slight beginnings, and in some instances from different activities. For example, Caleb Marshall, born in Chester County, Pennsylvania, in 1805 or 1806, was involved in grain milling before he set up the Marshallton Iron Works in Newcastle County, Delaware. In 1856, he moved to Philadelphia, building the Penn Treaty Works in Kensington, which he and his sons gradually extended.[3] The sheet interests of another family provided an example not only of long-term growth and product evolution but also of even more complicated geographical patterns in the foundations for the sheet mill industry of the early twentieth century. After some 20 years of effort, W. D. Wood of McKeesport Iron Works, near Pittsburgh, succeeded in the mid-1870s in producing Russian sheet iron or "planished sheet"—a product with a particularly smooth and flat surface. A quarter century later the Wood interests planned a Bessemer steelworks ten miles further up the Monongahela. A third of the billets or sheet bars made there were to be delivered to associated mills, at McKeesport, to the Wellsville Plate and Sheet Iron Company works on the upper Ohio and even to the Alan Wood Company and J. Wood Brothers of Conshohocken, on the west side of Philadelphia.[4] By the early 1890s, there was already an Association of Iron and Steel Sheet Manufacturers, meeting monthly in Pittsburgh. Later estimates suggest that between 1890 and 1900 production of sheet of 13 gauge or lighter increased from 300,000 to 700,000 net tons—or from less than 4.5 percent to 6.6 percent of all rolled iron and steel products.

The growth of the tinplate industry, in effect a wholly new creation of the 1890s, provided another boost to the sheet trade. From the late 1870s, tinplate manufacture in the Old World moved from the use of sheet iron to steel sheets (commonly called black plate, an uncoated plate) in the dipping plants which produced the finished product. As a result, though serving different end uses, throughout their histories the sheet and tinplate industries were intimately associated. The first American canneries were located in the East but shortly after the Civil War, with the westward movement of agriculture and the development of fisheries, they began to appear in the Midwest and California. Alaska's first cannery dated from 1878. By the end of the century output of canned foods had already reached 900,000 long tons of net contents.

Long before the end of the nineteenth century, the United States was far and away the world's largest market for tinplate, its annual consumption increasing by about 300 percent between 1870 and 1890. Until the latter date, all of the tinplate was imported, overwhelmingly from South Wales. Despite the fact that it was rapidly becoming the world leader in steel, time and again attempts to produce American tinplate failed. Early problems of a lack of skilled labor, unsuitable machinery and poor methods could have been overcome. But sheet and tinplate mills were labor intensive as compared with most of the rest of the industry, and the relatively low wages in Wales and carefully managed competitive pricing by producers there frustrated every effort to make this product profitably in the United States. Even so, repeated attempts were made to force an entry into this trade.

In 1858–1859 tinplate was made for a short time by the Hussey Copper Works in Pittsburgh. A decade later the Cambria Iron Company and the Pittsburgh quality steelmaker James Park sent men to Europe to learn the necessary techniques. It is said they returned convinced that, while America could offer equal or even superior facilities in other respects, because of high wages it could not compete with the Welsh without increased tariff protection. As tinplate prices increased in the early 1870s three more companies in the rising western centers of the iron and steel industry tried to make tinplate—at Leechburg and Demmler in western Pennsylvania and in Wellsville, Ohio. Each made a good product, but after a short time succumbed as Welsh producers and tinplate importing houses in the eastern United States, recognizing the threat to their livelihoods, cut prices to levels at which they were unprofitable. W. C. Cronemeyer, the so-called father of the American tinplate industry, provided a short but colorful account of the work of the agents who carried out this process. At intervals over the next decade, Demmler turned out small amounts of either tinplate or terneplate (sheet coated with a lead-tin alloy) in addition to its fine sheet iron and steel.[5] But for all practical purposes the United States continued to depend wholly on Wales. At last federal action provided an escape.

The McKinley Tariff Act of October 1890 provided for an increased duty on tinplate of 2.2 cents a pound, to come into force in July 1891. The response was immediate. During that year some 20 tinplate works were brought into production, and another ten were reported as under construction.[6] Some of them bought their supplies of black plate, others supplied their own. By

February 1892, of thirteen works built or building in Pennsylvania, five were intending to buy black plate, six would make their own and the other two, though at present buying plate, planned to make it in the course of 1892. In the case of the Falcon Iron and Nail company of Niles, Ohio, the company at the beginning of 1892 was preparing to erect a complete tinplate works, the black plates to be supplied by sheet mills it had installed as early as 1864.[7] The Wilson Tariff Act of 1894 reduced the duty on imported tinplate to only 1.2 cents per pound, but by then a domestic industry had been firmly established, output was rising at an increasing rate and many new works had been built. The industry now included some bigger plants, and it was usual for the companies to make their own black plate. To take one example: the plant of the Beaver Tin Plate Company of New Lisbon, Ohio, begun in October 1894, by the following April was expected to have in operation a rolling mill department of four hot and four cold mills with an annual capacity of 6,000 tons of black plate. It would produce both tin- and terneplate. Its president was C. W. Bray and the secretary I. M. Scott. The former would later play a significant part in developing sheet rolling technology, the latter in the business organization of the sheet and tinplate industry.[8]

In 1890, all tinplate was imported; seven years later, home output was two and a half times as large as the amount imported from Wales. From only 1,118 tons in its first, part year, within nine more output reached 425,002 tons. Almost inevitably there was overexpansion of plant. By 1901/1902 the aggregate capacity of the tinplate mills was reckoned to be 700,000 tons, but 1900–1903 output averaged only 453,000 net tons. Almost from the start, the industry was volatile, some operations disappearing almost before they were fully established. An extreme instance was that of the Locust Point Iron and Steel Works, near Baltimore. Early in 1892 it was named as preparing to enter the trade, with a complete tinning plant and its own black plate. By the fall of 1893, it was in the hands of receivers.[9]

The new trade had an immediate impact on the sheet mills. In 1891–1892, almost a third of the black plate tinned in the USA was brought in from Wales. Now some long-established sheet firms began to offer black sheets for tinning. At least 5 sheet mills which existed in 1888 were also making black plate ten years later. Both new tinplate firms and independent companies built sheet mills to supply the new trade.

For 1898, capacity figures are available for just two thirds of the total of sheet-making plants. Their average annual capacity was a little over 13,000 tons. The few works in the East were either smaller than this or, in the case of eastern Pennsylvania, of average size. Already there were large works west of the Appalachians. Vandergrift in western Pennsylvania was the largest, with 89,000 tons sheet capacity. Elwood, Indiana, was capable of 65,000 tons and the recently built Shenango works could make 60,000 tons. The combined capacity of all mills for which figures are available was just over 1 million tons. Even if one assumes that plants for which no figures were cited were likely to be the smaller ones, the total capacity of sheet mills must have been about 1.25 million tons. National output of tinplate averaged 400,000 tons in 1898–1900, and the total output of sheet of 13 gauge or lighter in 1900 was later estimated at 700,000 tons. In short, the surge in investment in new plant meant that sheet rolling capacity like that for tinplate production was overextended. Such a situation induced companies to cut costs so as to increase their market share. It also provided an incentive for combination and rationalization.

Concentration on fewer, larger plants was one of the leading industrial trends of the age, but it was less obvious here than for most finished products. This was in extreme contrast with the situation in Bessemer rails. For that bulk, relatively low-value product, it was essential that a rolling mill should have supporting steel capacity. Except for the Springfield mill, a small and not really competitive operation, all rail-making plants also had blast furnaces. Sheet was produced in a greater variety of qualities, but in small average tonnages. Rail mills pioneered innovations and mechanized processes wherever possible so as to cut costs, but sheet rolling remained labor intensive. Its finished product commanded a much higher price. In 1898, the average price for standard rails was $17.63/ton, and in 1899 $28.13; in the same years, black sheets averaged $42.28 and $60.24. In 1898, six works making "special" steel sheets could call on their own crucible steel capacity. Of the remaining 108 sheet works six only had steel plants. Even so, this mere handful of steel/sheet works accounted for roughly one quarter of the total capacity in this sector of the industry. Other mills depended on purchased sheet bars. Only two of the sheet works had blast furnaces.

Considered from the point of view of producing districts, two thirds of the known rated capacity in 1898 was focused in five small areas. In four of them, annual capacity for sheet was between 140 and 150 thousand tons. Although it had the largest number of plants for which no details were available, it seems probable that the plants centered on Pittsburgh, but spreading out to other parts of Allegheny County and nearby parts of Washington County, made up the largest concentration of sheet capacity. The average size of the ten plants in this area for which details are available was 14,200 tons. For six plants in the Mahoning and Shenango valleys, the average capacity was 24,000 tons, in six operations in a small area of central Indiana 20,500, and the eight mills in the Kiskiminetas valley (in 1888 there had been only two) east of Pittsburgh averaged 18,000 tons. Seven plants along a short section of the upper Ohio from Wheeling northward had a combined 1898 capacity of

Table 1.1. The sheet and Bessemer rail trades, 1888–1901 (capacities in thousands of tons)

	Sheet		*Bessemer Rails*		
	1888	*1898*	*1890*	*1898*	*1901*
Number of works	75	114	13	13	11
Annual capacity	n.a	1,004	2,650	3,170	4,630
Av. Cap per works	n.a	13.2	203.8	243.8	420.9
Av. Cap. 3 largest works	n.a	71.3	350	420	650
No. of works with steel cap.	n.a	6?	13	13	11
No. of works with b.furnaces	n.a	2/3?	12	12	11

Source: Knox 1943; Historical Statistics of the United States

103,000 tons or a works average of 13,000 tons, but the largest three operations in this area, each in Wheeling, averaged almost 23,000 tons.

The reasons for concentration in these five areas differed. In and around Pittsburgh, development of sheet and tinplate capacity reflected the diversifying activities of the nation's largest, longest-established rolling mill center. By contrast, in the "Valleys" sheets and tinplate were taken up as currently expanding lines by producers who were venturing into the steel and rolling mill businesses for the first time. For some firms in Wheeling and in nearby Ohio valley localities, investment in sheet mills and tinplate provided a hopeful new opportunity for those whose traditional cut nail industry had withered under competition from wire nails. Although it benefited from its relative proximity to the meatpacking industry, the Indiana industry was above all an opportunist response to a sudden surge in natural gas output there following its discovery in 1886. Communities eager for employment made almost free-gift offers of this new source of energy to the rolling mills. Mills along the Kiskiminetas River were an unusually specialized group.

Apart from the five major concentrations of capacity, large numbers of individual plants were scattered through country districts, particularly in Pennsylvania and eastern Ohio. As with the Indiana gas belt plants, many originated in responses to tempting offers from this or that small local community, eager to climb on a perceived bandwagon of economic growth. Cases in point were Piqua and Cambridge, Ohio, the latter of which had two sheet works by 1898. Another instance, whose local initiative and company response may be traced, was that of Dennison, Ohio. There seemed every expectation of mutual benefit from this form of promotion—favorable terms for the company; jobs, income and growth for the community. In many cases things worked out well, a boom in local population figures indicating the progress achieved. But occasionally combination into bigger groups brought a sudden closure of the local plant. Even if this was avoided, such relatively isolated mill towns, often almost one-industry settlements, would prove vulnerable if anything led to new processes requiring concentration of production.

By 1900, sheet, black plate and tinplate works were widely but unevenly spread across the manufacturing belt of the northeastern states, from New York City and Baltimore through to St. Louis. The distribution pattern within this area was to be of enduring importance, for it formed the geographical foundation onto which the revolutionary rolling mill technologies of the new century would be grafted.

NOTES

1. John P. Lesley 1859, passim.
2. *BAISA* 19 December 1888.
3. *BAISA* 9 July 1874, 1 March 1898.
4. Knox 1943, 47.
5. W. C. Cronemeyer, in *BAISA* 17 October 1888.
6. W. C. Cronemeyer, quoted *BAISA* 20 August 1895.
7. Knox in *IA* 18 November 1943, 469.
8. Jeans 1902 p 163; Knox, 31.
9. *BAISA* 10 February 1892, 1/8 November 1893.

Chapter Two

Growth, Combination, and Rationalization, 1898–1901

In the 1890s, sheet production became a large industry, meeting increasing demands from a multitude of expanding consuming industries. A tinplate industry had been created not so much because of an immediate rise in demand but as tariff reform and entrepreneurship together squeezed out a traditional, monopoly, external supplier. Until 1897, home consumption of tinplate showed no strong tendency to expansion, but over the next three years there occurred a sustained and major increase. Price trends contrasted with these changes in output. After rising in the early 1890s, from a high in 1895 the annual average price for tinplate fell for the next five years. The cause for these different trends was a repetition of a long-established feature of the steel industry—an unduly rapid increase in capacity reflecting overambitious plans by a large number of entrepreneurs. It was a characteristic particularly marked in this product because there were a large number of separate decision makers operating relatively small business units in which costs of entry or expansion were low. It became accepted dogma that the only remedy for such unsatisfactory commercial conditions was combination followed by a weeding out of inefficient plants and, if possible, restraint on the building of new ones. Individual units of production might remain small but would be combined in larger associations of capacity and control. Overexpansion and combination alike were to define the conditions in which companies, plants and technologies would operate in the early years of the next century.

During the four or five years after the passing of the McKinley Tariff, headlong expansion in tinplate was associated with an anarchic process of growth. Rolling mills to produce black plate were of relatively small capacity, and plants and could be set up cheaply by small companies. Both the incremental expansion of existing mills and the entry of newcomers to the trade were easy and cheap. By January 1892, only six months after the McKinley tariff came into effect, 20 works were already either in operation or prepared to make tin- or terneplate. Five years later, there were 69 completed works, as well as one work being rebuilt and four more in course of erection. As early as summer 1894, the Iron and Steel Association published a note headed "Tinplates Galore!"[1] It was a time of low prices in steel generally, but over the next two years as most prices strengthened—including that for the semi-finished steel used by sheet and tin mills—tinplate prices continued to fall. In spite of this, more capacity was built, including wholly new plants. In 1896 works were built at Connellsville and Washington in western Pennsylvania, in 1897 black plate mills were installed at West Newton on the Youghiogheny and at Dennison, Ohio, and the National Tin Plate Company put up a major works at Monessen on the Monongahela. Next year there were new mills for Johnstown and at Newcastle in the Shenango Valley. In the latter area, 60,000 tons of sheet for tinning capacity was installed by the Shenango Valley Steel Company. It became obvious that without some form of coordination prices might continue to fall.

As early as March 1897, tinplate makers met in Monongahela House, Pittsburgh, to consider forming an organization to regulate the distribution of tinplate so as to ensure a reasonable profit. Six men were appointed as a committee to produce a prospectus for a new association. In the meantime, a temporary agreement was drawn up, binding on all so long as it was agreed by mills with at least 90 percent of the capacity.[2] Consumption increased, but the fall in prices continued, in 1898 to as low as $2.99 per base box. Then, in December that year, the trust-building Moore brothers of Chicago managed to put together a wide association, the American Tin Plate Company. This included 38 plants, some two thirds of the industry total, with approximately 280 mills. Works were purchased not with cash but by payment in the stock of the new combine. Nominal capital was $50 million, but only $40 million was issued. A historian of the industry later wrote that "What became of the other $10 million worth of stock is a matter of conjecture."[3] (In fact, a year after the merger, Daniel G. Reid, first president of American Tin Plate, revealed to the Industrial Commission that the "surplus" $10 million had been assigned to Judge Moore "to cover all the expenses of bringing the companies together and perfecting the organization of the syndicate. No control was exercised over Judge Moore as to the disposition of this stock, which was put into his hands to do with as he thought fit.")[4] The consolidation succeeded in checking the tendency for prices to fall and by early autumn 1899 tinplate prices had been increased to $4.65 per base box. The 1898 earnings of the constituent companies of American Tin Plate had amounted to $2.0 million; in 1899 they rose to $3.60 million and in 1900 were $5.75 million.[5]

Table 2.1. Tinplate in the United States, 1890–1900 (thousand net tons)

	Consumption	*Production*	*Imports*	*Price per box*
1891	368	1	517	$5.34
1892	321	21	211	$5.30
1893	345	62	314	$5.37
1894	324	83	227	$4.89
1895	373	127	253	$3.87
1896	314	180	192	$3.63
1897	381	287	115	$3.26
1898	441	366	85	c.$3.00
1899	470	404	54	$4.41?
1900	493	425	74	$4.82?

Source: Knox 1943; Historical Statistics of the United States

The new company tried a number of ways to cut costs, in the process taking some actions which, to a more fastidious generation, may seem near to restraint of trade or worse. Selling was centralized, removing the need for the men who since the heyday of imports from Wales had handled the requirements of tin can makers. More system was introduced into rolling programs—all orders were to be sent to Chicago to be allocated to individual works. The combine used its size to get a reduction in the price it paid for steel. Then, in order to help improve raw material supply instead of continuing to depend on purchases from major outsiders such as Federal Steel or Carnegie Steel, the Moore group formed the National Steel Company from a number of small steelworks and bar mills. It set out to make its own requirements more efficiently. Special arrangements were also made with those accounting for some two thirds of the supply of mill rolls—although the combine's president later claimed he could not remember whether they had also contracted with those suppliers to ensure they did not also provide equipment to non–American Tin Plate mills.[6] Independent dippers now found it more difficult to buy black plate, and some had to bear the high cost of importing this over the tariff: in 1899 they held a meeting at which they considered starting a mill of their own.[7] Even more dramatically, what were assessed by American Tin Plate to be inefficient or badly located plants were marked out for closure and the removal of their machinery. Unfortunately for its success, action of this sort caused the layoff of skilled engineers and managers at the same time that prices were rising. In such circumstances, many experienced men chose to reenter the trade on their own account.

By summer 1899, American Tin Plate was already dismantling some peripheral or otherwise unsuitable works. These included a works in Brooklyn, two in Baltimore (a third Maryland works, in Cumberland, was closed in

1901/2) and the Britton Rolling Mill of Cleveland. By October Reid disclosed that 12 or 13 mills had been closed because they could not operate "to advantage." As another step toward concentration, the West Newton plant was dismantled, removed and consolidated with a works in Connellsville. The Blairsville mill was idled at the same time; it was expected it might never run again. (It was abandoned in 1901/1902.) In 1900, the Montpelier, Indiana, plant was abandoned. About 25 new mills were opened in various locations.[8]

Efforts were made to ensure those who had sold plants to American Tin Plate did not reenter the trade. There was also an arrangement stipulating that the previous owner should not again engage in tinplate production within 1,500 miles of the old plant for at least 15 years. The inadequacy of such a scheme in these times of huge consolidations, sweeping programs of rationalization and shaky business ethics was spectacularly exposed years later in the Dissolution Suit against the United States Steel Corporation. After selling his New Castle Wire Company to John Gates, John Stevenson organized the Shenango Valley Steel Company of Newcastle in 1891–1892. By 1898, 60,000 tons of sheet for tinning capacity was under construction there, when its black plate mills—but not its steel plant—was sold to the American Tin Plate Company. The following year, Stevenson sold the Newcastle steelworks to National Steel, receiving $3.3 million for what was capitalized at only $1 million. Encouraging his business associates with the homely philosophy, "We'll shake the apple tree again," Stevenson joined with F. H. Buhl to organize the Sharon Steel Company and in 1901–1902 established the Sharon Tin Plate Company.[9] Except for its shameless brashness, Stevenson's case was not unique. Independent firms continued to be attracted to the trade. As a result, although by the time of the 1901 edition of the AISA Works Directory the number of completed tinplate works had been reduced to 55 as compared with 69 five years earlier, seven new plants were being built and another was projected.

The sheet steel industry was bigger and older than the tinplate trade, but here too expansion in the nineties made combination desirable. In March 1900 an American Sheet Steel company combined 25 previously independent firms, controlling about 70 percent of national sheet capacity. As with American Tin Plate, with which it shared promoters, the nominal capital of the sheet combine was $50 million. Unlike American Tin Plate, it had some steel capacity of its own, mainly at Vandergrift. In 1900 it made about 150,000 tons of steel.[10] It began to buy much of the rest of the steel it required from National Steel. Attempts to reduce costs involved arrangements with some of its competitors. When the Republic Iron and Steel Company was established as an amalgamation of rolling mill companies, its capacity included 15,000 tons of sheets. A year later American Sheet Steel came to an agreement to withdraw from the merchant bar trade in exchange for Republic's withdrawal from rolling sheet. There was some exchange of plants.[11]

In spring 1901, both American Tin Plate and American Sheet Steel became subsidiaries of the United States Steel Corporation. Most immediate control and decision-making remained unchanged and rationalization continued. Between spring 1901 and December 1902, the Tin Plate company abandoned ten works, and by the latter date had only 23 works with 250 mills, 11 percent fewer than at its formation four years before. Among units closed in this period was Blairsville Rolling Mill and Tin Plate works built in 1892 and the "new" works Caleb Marshall had erected in Philadelphia in 1856. Altogether, by 1905 of the 40 works with 279 mills which had entered American Tin Plate in December 1898, it had closed 24 containing 98 mills. Of 29 sheetworks brought into US Steel, American Sheet Steel closed five straightway or very shortly after incorporation. The Apollo Iron and Steel Company's sheet mills in Apollo, Pennsylvania, were written off in 1903, but Vandergrift continued to flourish under US Steel. It provided an early illustration of what a generation later became notorious, the power of corporate decisions to make or unmake communities. Between 1900 and 1910 the population of Vandergrift and Vandergrift Heights increased from 3,986 to 7,314; Apollo did little more than hold the line, with 2,924 and 3,006 respectively.[12]

One of the casualties of rationalization was the works which the citizens of Dennison, Ohio, had helped secure for their town in 1897. The mill last operated in 1902, but it proved difficult to eliminate it from the establishment of the company. On 13 December 1904, the executive committee of American Sheet and Tin Plate, a new, combined operation within United States Steel, held a meeting in the offices of Judge Gary at 71 Broadway. The sharpness of disagreement exposed was not fully hidden by the measured, even stilted minutes of this meeting, for the life of a community was at stake: "Whereas at the formation of this company there was transferred to it by Dr Wiley Caughy, together with certain other properties, a property known as the Dennison Rolling Mill, which mill had been erected upon property donated by the citizens of Dennison, Ohio, and also in part with funds also donated by said citizens, and made subject to certain conditions for continuous operation, and whereas it has been found by the company that due to its location and equipment, it could not be operated with sufficient economy to justify such operation, and due to suit brought by said citizens for the enforcement of the condition referred to, it was deemed best by the officers of this Company to compromise the question involved in said suit, by deeding back to the said citizens the land and buildings in question." This action was ratified by the meeting.[13] The Dennison mill was written off by US Steel in 1905.

In 1903, the two subsidiaries of United States Steel, American Tinplate and American Sheet Steel, were amalgamated as the American Sheet and Tin Plate Company. A few months later this internal combination was reckoned by a US Steel director to cut annual operating costs by $250,000. The new subsidiary continued the process of concentrating its plants. In December 1903, the Tinplate company had 20 works and the Sheet company had 19, but by 1910 American Sheet and Tin Plate owned only 33 works. But concentration of capacity and other cost savings could not prevent new producers from springing up. National output of thin flat-rolled products was risin, and sheet and tinplate plant was relatively cheap to install and required less supporting or ancillary plant than most steel products. Independent firms increased their share of output more than in the whole steel industry. Between 1900 and 1905, the proportion of all finished rolled steel controlled by other companies than US Steel rose from 48.7 percent to 51.9 percent, but in plates and sheets from 40.6 to 50.2 percent. In tin- and terneplate, the advance was even more impressive—from 26.6 to 39.0 percent.[14] Many firms and works later of great importance originated at this time as small enterprises in what was almost a frenzy of new company creation.

By the end of 1901, of the 55 tin- and terneplate works in the United States, USS controlled 25. Seven more works were being built and another was projected, only one of which belonged to US Steel.[15] Some plants built at this time or shortly afterward remained relatively small factors in the future shape of the sheet and tinplate industry; others eventually became major, fully integrated, multi-plant operations. An example of the first was Follansbee Brothers. For more than 20 years, W. U. Follansbee worked for a Pittsburgh tinplate importing and jobbing firm. He lobbied for a duty to permit the United States to build its own industry, but not until the creation of US Steel helped bring more stability to prices did he enter production. Buying a 250-acre farm along the Ohio in Brooke County, West Virginia, he built a sheet mill there, then an open-hearth plant to supply it and afterward tin- and terneplate plant

Table 2.2. Tinplate works in the United States, 1891–1903

	Completed	*Building*	*Projected / rebuilding*
1891 Ap.	3+	2	n.a
1892 early	20	10	n.a
1895 end	69	4	2
1898 Ap.	69	1	n.a
1901 end	55	7	1
1903 end	53	2	1

+ Plus one works making terneplate
Sources: AISA; Jeans 1902, p. 163

to process its output.[16] Eventually he built a second plant. Follansbee operations never became a major factor in sheet steel.

NOTES

1. Preface to the 13th Directory of the Iron and Steel works of the United States, 1896, quoted *BAISA* 1 March 1896, 53, BAISA 8 August 1894.
2. *BAISA* 10 March 1897.
3. Knox 1943, 22.
4. *IA* October 1899, report of the proceedings of the Industrial Commission.
5. Craig 1966.
6. D. G. Reid to US Steel Corporation Dissolution Suit, quoted *IA* 23 May 1912, 1320.
7. *IA* 16 May 1912, 1242.
8. *BAISA* 1 November 1899; IA October 1899, 16 May 1912, 1242, 23 May 1912, 1320.
9. *IA* 30 May 1912, 1349, 1350.
10. C. M. Schwab to A. Carnegie 24 January 1901. ACLC.
11. *IA* 6 April 1899; Clark Vol. 3, 88, 89.
12. USS Annual Report for 1902; list of abandoned properties by USS Property Bureau 29, May 1946.
13. Mins of Executive Committee AS&T, 13 December 1904.
14. Jones 1922.
15. Jeans 1902, 161.
16. "WU Follansbee" in *National Cyclopedia of American Biography* 29, 1941; BFSP March 1941, 313, 316.

Chapter Three

Vandergrift Works

Until the early 1890s, there seems to have been little distinctive about the design and practice of American sheet mills. Essentially they were like those in Great Britain, whose main center was the Black Country of Staffordshire. In tinplate, much of the equipment used in the early mills was Welsh. By the early twentieth century, methods of production in the United States had become so generally superior that United Kingdom visitors went home recommending modernization on American lines. In fall 1899, Daniel Reid claimed that American tinplate mills had already about 30 percent greater output than foreign mills. Three years later, a British merchant, Alfred Mosely, personally financed an industrial commission made up of leading representatives of workingmen to study United States productivity. The reports of the Mosely commissioners on the steel industry were disturbing in their home country. James Cox, of the Associated Iron and Steel Workers of Great Britain, was impressed with what Americans had done with Old World technology. He reckoned methods in British sheet mills were poorer, though in pointing to manufacture of their own steel as an important factor in the success of American mills, he failed to recognize that the Vandergrift mills which he eulogized were also exceptional in this respect. Surprisingly he thought American sheet mills had little to show in saving labor, but he acknowledged they were very superior in general layout. Black plate mills in the new American tinplate industry had strong, well-powered mill trains, contrasting with "the feeble and spasmodic grind at many of our own works." "Throughout their pickling, annealing, tinning and finishing processes, they are very much ahead of us in the nature of appliances and methods of work." American works were relatively free of smoke. At home, "the tinplate trade, like the sheet trade, is over weighted with obsolete machinery and the incubus of poverty."[1] A few years later, another Englishman, this time a newly arrived immigrant, was surprised to find by how much the sheet industry in which he took employment had outdistanced that in his native land. Writing from Cleveland over half a century later, he referred to this contrast: "Between the years 1875 to 1900 the sheet mills in this country had made great advances both in the general layout of the plants, and the rolling mills themselves, much sturdier than anything we had in Great Britain and also in the quality of the sheet, and also in galvanizing, and corrugating and pickling, annealing and cold rolling."[2]

An important part of the sheet industry was located along the Kiskiminetas River, on the boundary between Armstrong and Westmoreland counties, in western Pennsylvania. Here was the largest plant and some of the best practice. In the second quarter of the nineteenth century, David Laufman and Thomas Chambers of Chambersburg in east central Pennsylvania had built blast furnaces on South Mountain. Their iron had been hauled to Carlisle to be hammered into shapes. In 1840, a son, P. H. Laufman, moved westward to work in the Pittsburgh hardware business. In 1876 he bought an old rolling mill at Apollo in the Kiskiminetas valley 35 miles from Pittsburgh, and two years later began experimental production of tinplate. His early efforts failed, but Apollo became a mill center, with an 1882 capacity of some 8,000 tons of fine sheet a year. Later the Laufman company opened sheet mills at Freeport and at Saltsburg.[3] In 1894, the shipping clerk for PH Laufman at Apollo was a young local man of 21. Six years later, when his company was absorbed into the newly formed American Sheet Steel Company, he was appointed superintendent at Apollo. Many years afterward he, William A. Irvin, would play an important part in the national evolution of the sheet and strip mill industry. At the end of the nineteenth century a more prominent place was occupied by the owner of another Apollo venture.

In 1855 the Kiskiminetas Iron Company was formed to mine coal and iron ore, roll iron and make nails and spikes. It changed hands a number of times. By 1874 it was operating seven puddling furnaces and two sheet mills. In 1882 a post–Civil War immigrant from Ireland, George McMurtry, took over this works as the Apollo Iron Works, by now a small galvanized sheet plant. Two years later it became the Apollo Iron and Steel Company organized by a group which included McMurtry and an enterprising 59-year-old, Jacob J. Vandergrift. (Earlier Vandergrift had been a riverboat captain, had moved into oil refining, operated one of the first natural gas pipelines and was involved in banking.) The Apollo mills were rebuilt, and acid open-hearth furnaces were added. By 1898, when Laufman's Apollo works had a capacity of 6,000 tons of fine and black sheet, the nearby Apollo Iron and Steel Company was capable of 14,000 tons a year of sheet for galvanizing.[4] Even before that, it had built a much larger works on a 640-acre tract, until then occupied by farmers, and located some three miles down the valley.

McMurtry staunchly refused to negotiate with unionized labor and consequently had to contend with over a year of strike action including violence from October 1893. By 1895, he had decided to build a new mill in a place which would be provided with a planned environment for its workers. The new town and works were given the name "Vandergrift." It was, as Ingham put it, "a classic example of the new welfare capitalism." Some described it as a model town, another referred to its "near-Utopian conditions," but it was almost a one-industry community.[5] To provide jobs for those who lived in his new town, McMurtry built an exceptional plant. Its first sheet was rolled in October 1895, and between then and 1898 this works of the new Apollo Iron and Steel Company was provided with six open-hearth furnaces, a blooming mill, a continuous sheet bar mill (each of these units designed and constructed by the eminent steel engineer, Julian Kennedy) and 21 hot sheet mills. By the time the American Iron and Steel Association conducted its survey of iron and steelworks in 1898, the parent works at Apollo had a small steel plant and an annual capacity of 14,000 tons of sheet for galvanizing, but Vandergrift could make 70,000 gross tons of steel, 96,000 tons of sheet bar, 38,000 tons of black sheet and 51,000 tons of sheet for galvanizing.[6]

By 1901, Vandergrift was reckoned the world's biggest sheet-making plant. More important, it was outstanding for efficiency. Although its blooming mill was criticized by two steel men from Middlesbrough, England, in 1899, most visitors were impressed.[7] When in late summer 1900, the Carnegie Steel Company considered installing sheet mills at Duquesne, the board was told that the Vandergrift mills had the most modern plant.[8] By 1901/02 weekly output of galvanized sheet was as much as 2,000 tons. A large extension was underway, and management was reported as to have in view a plant of possibly as many as 50 mills.[9] By 1910 there were 10 open-hearth furnaces and 37 sheet and jobbing mills. Two years later the Englishman cited above, who had worked for many years in the Staffordshire sheet trade, arriving in western Pennsylvania, took a job at Vandergrift. He later recalled his impressions: "I can assure you that when I emigrated to this country in 1912, I left England with the fixed impression that the Orb Iron Works of John Lysaght Ltd. in Newport, Monmouthshire, with its 44 mill stands and its product, was just about the finest plant in the world. However a few months in the Vandergrift plant of the old American Sheet and Tin Plate Co., with its 35 mills and its product of all grades (no tinplate however) and its pickling, annealing, cold rolling, galvanizing etc. and above all its transportation of material around the plant, and its wonderful loading and shipping facilities and above all else, its much greater tonnage output per 8 hour shift, all this completely changed my outlook. My old country was far behind."[10]

The size, facilities and quality of the Vandergrift mills were exceptional, but even so its longer-term, marginal status had effectively been decided before World War I. There were two main problems, external and internal. Early in 1906, US Steel decided on the construction of a completely new, fully integrated steelworks on the Great Lakes shore at Gary, Indiana. By February 1909, Gary was making iron, open-hearth steel and rails. On 16 November 1909, the US Steel finance committee took another important step, considering estimates for providing Gary with sheet and tin mills to be built by AS&T, which had not been included in the original plans for the complex. Construction of sheet mills began in 1910.[11] Compared with this new plant, built on virgin, flat land, which, if necessary, could be extended by infilling nearby parts of Lake Michigan, Vandergrift occupied a relatively small site and was confined within the meander of a small river. It depended on cold metal practice, and its open-hearth furnaces used scrap or bought in supplies of pig iron. In 1917, it seemed for a time that a way out of the last of these restrictions might at last be found. A careful study was undertaken by the works assistant manager, H. L. Bodwell, and Joseph Malborn for a new works, integrated from blast furnaces to finishing mills to be located at Hyde Park along the Kiskiminetas three miles north west of Vandergrift. No development followed their report.[12] By 1922, the 192,000 tons of sheet and light plate capacity at Vandergrift was far bigger than the thin flat-rolled tonnage at any of the long-established AS&T works, but Gary now had an annual capacity of 332,000 tons of sheet and light plate and 186,000 tons of black plate for tinning.[13] Logically, when a radical new technology was taken up no more than five years later by AS&T, it was installed not at Vandergrift but at Gary.

NOTES

1. *Report of Moseley Industrial Commission to the United States*. October to December 1903, 44–47; see also E. Parkes in Jeans 1902, 565.
2. J. Malborn c. 1956, and in correspondence with the author 1966.
3. *BAISA* 6 December 1882; *IA* 6 May 1926, 1314.
4. AISA *Iron and Steel Works Directory* 1898.
5. *IA* 17 June 1897, 7; J. Ingham, 137, 138; *BAISA* 15 September 1912, 84.
6. AISA *Iron and Steel Works Directory* 1898; USS "A Brief History of Vandergrift Plant"; *Iron and Steel Engineer* October 1933, xiv.
7. W. H. Parker and C. L. Bell, *Report on the Steel works and Rolling Mills of the United States*, Middlesbrough 1899, 6, 21.
8. Carnegie Steel Company. *Mins of Operating Department* 23 July 1900 ACLC 77.

9. Jeans 1902, 305, 565, 569–72.
10. J. Malborn to author 1966.
11. USS Finance Committee 16 November 1909.
12. J. Malborn to author 1966.
13. *IA* 1 January 1925. (The statistics relate to December 1922.)

Chapter Four

Expansion and Company Promotion, 1900–1920

By the early twentieth century, expansion in the manufacture of finished iron and steel sheets, along with the recent surge in tinplate production, and its consumption of black plate, had caused a rapid expansion in sheet-rolling capacity. In 1901, J. S. Jeans, secretary of the British Iron Trade Association, made an extended visit to iron- and steel-producing centers in the United States. He noted great expansion then underway in sheet, "for which the United States appears to have a most devouring appetite." By the end of that year, the United States Steel Corporation was extending some of the works it now controlled through the American Sheet Steel Company. Thirty-one independent companies were reported to be building new works or extending existing ones.[1] This program proved to be only the beginning of major extensions. Between 1900 and 1914, output of tinplate and of sheet steel roughly doubled. For lighter grades of sheet (13 gauge or lighter), production of which was only separately listed from that of plate from 1905 (before which there were only estimates), the advance was even more rapid. The large increase was accompanied by further advances in technology and a reaching out toward even more radical changes. Yet more companies were established, some of them to play key roles in the industry for decades to come. There were major new focuses of sheet-making capacity.

Until well into the new century, growth in sheet output was closely linked to expansion in well-established outlets. One account of 1906 describing the rapid increase mentioned heavy demand in making steel railcars—presumably it was heavy-gauge material—another listed metal lathing, increasingly used in housing, and sheet used in producing cases for bottles—"particularly beer bottles." Consumption of electrical-grade sheet, negligible only a few years before, was now 80–100,000 gross tons a year. Sheet was being used in garbage cans, and the large bilge barrel was important as a

Table 4.1. Production of light sheets, 1890–1920 (thousand net tons)

1890	300	1916	2,520
1900	700	1917	2,651
1905	983	1918	2,322
1909	1,249	1919	2,352
1914	1,302	1920	3,232

Note: Figures for the earlier years are approximations

consumer of extra-deep drawing sheets—in heavier gauges.[2] Early accounts of growing markets failed to mention the one which over the next few years would rapidly increase in influence and before long became the leading single factor—automobile production.

In this period, manufacture of automobiles began to affect various parts of the steel industry. At first improvisation was common—crankshafts were made from billets intended for rails, gears from tool steel and cylinders from iron suitable for stoves.[3] The most important impact of the new industry was in sheet steel, for this material gradually replaced the mainly wooden materials used in making the traditional coach or carriage. Early car bodies were built in the same way as coach bodies, that is, by using panels of wood. When it was found that steel provided a better base for the application of paint, steel sheets were tacked on top of the wooden panels. The next step was to attach sheets directly to a wooden frame, using material of about 20 gauge. The sheets were then hand painted with up to 17 coats, rubbing down occurring between each coat. Under such conditions an auto body took up to 30 days to finish. Then, in 1923, Du Pont introduced a new lacquer, Duco, which could be sprayed on and dried quickly. Now only five to seven coats had to be applied, and the body could be completed in three days. This new treatment showed up the defects of the sheets used until then and therefore emphasized the need for an improved surface finish.[4]

As progress was made with the surface, the development of the enclosed car and the introduction of deep stamping of larger body shapes in quick-action presses emphasized the desirability of using deep-drawing sheets of uniform gauge throughout. E. G. Budd of Philadelphia, who made the first all-steel body in 1912, bought sheet on the basis of "understanding" rather than "specification"—that is, the body maker showed the mill firms what it expected the sheet to do and left them to adopt methods of production to meet his requirements. By 1927, Budd was using as much as 165,000 tons of sheets a year. In other words, after two decades of accelerating development, by early post–World War I years the automobile industry was providing sheet mills with quality requirements which within a few years would transform their industry.[5] The rapid rise in the number of automobiles made demanded greatly increased tonnages, thereby providing another incentive for rethinking methods of production in thin flat-rolled steel. In 1900, 4,192 passenger cars were made in the United States; in 1910 and 1920, the numbers were 181,000 and 1,905,560, respectively. Ford turned out 195 cars in 1903–1904, 19,900 by 1910, and in 1913 180,000 of all types. The Ford 1913 operations consumed 102,000 tons of steel.[6]

The sheet and tinplate industries were already heavily concentrated in Ohio and western Pennsylvania, and the strong localization of automobile production in southern Michigan confirmed the advantages of these relatively nearby areas. By 1904, Michigan accounted for 26.6 percent of the value of the nation's automobile output and by 1909 for 38.8 percent. Over the ten years to 1914, whereas the value of manufacturing output in Pittsburgh and Allegheny city increased from $211 to $246 million, in Detroit, manufacturing values at the same dates were $128 and $400 million.[7]

Although outpaced by sheet, tinplate consumption also increased substantially. Demand was 0.3 million tons in 1890 and 0.8 million tons by 1911; by the latter date, it was almost wholly home produced. Some estimates suggested that this independence had been secured at a high cost. The United States Steel Corporation dominated in both tinplate and sheet, but its share in both declined rapidly. According to its own figures, at its inception it controlled between two thirds and four fifths of the output of various thin flat-rolled products. Like the two previous combines, it continued to concentrate production. In 1903, American Tin Plate operated 19 works and American Sheet Steel 20. The capacity of the latter was 247,000 tons of open-hearth steel, 29,500 tons of puddled bars, 8,400 tons of charcoal blooms, and 193,000 tons of sheet bars. Finishing capacity was 53,000 tons of plain and black sheets and 160,000 tons of galvanized sheets. At the end of 1903 the two subsidiaries were merged as American Sheet and Tin Plate. By 1910, AS&T owned only 33 works. Over the next four years the number was reduced to 25. It constructed a limited amount of new plant.

US tinplate consumption increased from some 300,000 tons in 1890 to 800,000 by 1911. By then the tinplate industry had 25,000 workers, of whom 15 "independents" with a capital of $12 million employed 11,000 in 157 mills. At the end of 1902, USS had 164 operating sheet mills and 264 tinplate mills as compared with 136 and 71, respectively, for the independent companies. Within eight years, AS&T had 186 sheet mills and independents 169. A few years after the formation of American Sheet and Tin Plate, Carnegie Steel normally supplied it with on the order of 90,000 tons of bars a month. By October 1910, shipments were less than half that level. As the directors of Carnegie Steel then ruefully put it, "Of course their tin plate season is largely over, but they have not seemed to be able for some reason to hold their own against the independents in the price of sheets."[8]

Over the years, the relative importance of thin, flat-rolled products in the output of the United States Steel Corporation rose, but their share of the industry total fell. In 1902, the category "Sheets—black, galvanized, and tin plates" accounted for 9.56 percent of USS tonnage of rolled and other finished products for sale; by 1912, it was 12.06 percent. But, in sheet and tinplate, as in other finished products, the US Steel share fell and "independent" companies became more important. This was partly due to the cautious policies of the Corporation but in part to aggressive marketing and expansion by others. By 1910, US Steel accounted for 61 percent of national tinplate output and in sheets and plates (not then separated in its figures) for 48 percent.[9] After that the rate of its decline accelerated, and by 1920 its national share in their combined tonnage was little more than half that at its foundation.

At this time, some long-established steel companies played little or no part in the sheet and tinplate trades. This was the case at the beginning of the century with Jones and Laughlin, Pennsylvania Steel and its associated Maryland Steel, as well as at Cambria Iron and Lackawanna Steel. In 1909–1910 J&L erected its first tin mill, at the new Aliquippa works. Though revivified in this period by Charles M. Schwab, Bethlehem Steel did not make thin flat-rolled products until it bought the Baltimore Sheet and Tinplate Company in 1916.

Over the years, a large number of new sheet- and steel-making concerns were set up in the western part of the manufacturing belt. Without exception they were small ventures, involved modest outlays of capital, and were in early years dwarfed by the major integrated companies. Many depended on purchases of sheet bars—or the slightly smaller "tin plate bars"—from the bigger companies. This situation was by no means always satisfactory. For instance, early in 1908 US Steel was criticized by independents for selling sheet and tinplate bars to British mills for about $19/ton when it was charging home consumers $29. (The Corporation argued this kept its mills at work during a difficult time.) In 1905, Carnegie Steel decided to build a sheet

Table 4.2. Production of sheet and tinplate US Steel and all other companies, 1902, 1905, 1909, 1920 (thousand net tons)

	US Steel	*Others*	*US Steel %*
1902	622	403	60.7
1905	825	493	62.6
1909	1,025	909	53.0
1920	1,437	3,0851	31.8

Table 4.3. New sheet mills built or building, 1910

Youngstown Sheet &Tube	8	American Rolling Mill	10
West Pennsylvania Steel	3	Canton Sheet Steel	6
Thomas Steel	5	Follansbee Brothers	2
Massillon Rolling Mill	4	Phillips Sheet & Tin Plate	8
Stark Rolling Mill	1		

Source: ITR 5 January 1911

bar mill at its Columbus, Ohio, works. Other integrated firms also began to make sheet bars. In 1906 or 1907, Republic Iron and Steel converted an unprofitable rail mill into a sheet and tin bar mill, and before 1912 Corrigan McKinney had included sheet bar in the product range for its new Cleveland works. There were advantages in terms of security of supply for sheet makers to make their own bars, but many continued to be dependent on purchases.[10]

It was in this period that the Chicago area for the first time became an important factor in sheet and tinplate. Inland Steel was founded in 1893 by a Cincinnati scrap merchant, Joseph Block, along with his sons and a number of Chicago entrepreneurs—eight men in all—when they bought a rolling mill at Chicago Heights for $65,000. Eight years later, having decided to build new works, Inland chose a site 20 miles out of Chicago in the Calumet district, an area which, as a later Block recalled, was then largely bare of development other than "a railroad tower house surrounded by a vast expanse of seemingly useless sand." Fifty acres of this wasteland were purchased and on them Inland built four 50-ton open-hearth furnaces, a blooming mill, bar mill, eight sheet mills and a jobbing mill. Production of 400–500 tons of finished steel daily began in 1902. After 1910 structural, plate and rail mills were added, but from the start, Indiana Harbor had a footing in thin flat-rolled products.[11] By 1920, it was a major producer with a capacity for 150,000 tons of sheet, light plate and galvanized sheet. This was, however, far exceeded by the new US Steel plant at Gary, which by 1912 was the Corporation's largest operation for black plate for tinning and its second-biggest sheet mill installation. Eight years later, black and galvanized sheet capacity there was 370,000 tons, and by the end of 1922 the almost 520,000 tons capacity of the Gary Sheet and Tinplate mills ranked them far and away the largest plant in the whole sheet and tinplate industry.

The biggest share of expansion in thin flat-rolled products continued to be held by Ohio or nearby areas of neighboring states. The *Iron Trade Review* in its review of 1910 business contained a list of 47 new sheet mills, all in the western Pennsylvania–West Virginia–Ohio core of the industry.[12] Some had involved senior men displaced by changes in established companies. As in the past, others were sponsored or aided by localities eager to attract economic growth. Small-scale entry remained more possible than in most other steel lines. As late as 1912, Julian Kennedy estimated a tinplate mill could be built for as little as $500,000, a fraction of the cost for other major products.[13] One example of a small project involved the small town which had already seen its local sheet mills closed by US Steel in 1902 and whose bigger offshoot at Vandergrift had also passed to USS. In 1912, a new Apollo Steel Company was organized with $300,000 capital to build a 6-mill sheet works on a 9-acre plot, a site given by the local citizens. The works was expected to provide 500 jobs.[14] Sheet mills were built here during 1912 and 1913; steelmaking was added in 1916. By 1920 this works was being extended to a capacity of 108,000 tons of black and galvanized sheet. It failed to become a major factor in the trade, but a number of other companies established at this time did.

The Valleys district and the area around Canton and Massillon, almost 50 miles farther west, were both focuses of growth, aided by a complex series of mergers. In the mid-1890s, the Shenango Valley Steel Company of Newcastle and the Ohio Steel company in Youngstown had made sheet bar from their own Bessemer steel. In 1899, the Republic Iron and Steel had been formed as an amalgamation of ironworks and rolling mills. By means of an extensive rationalization program, it focused operations in the Valleys, in which it built up its steel capacity. Its main product was bars, but it also made sheet. Two officials, James A. Campbell and George D. Wick, left Republic in 1900 to launch their own venture. Fifty-five Youngstown-area citizens invested $600,000 in a plant which Campbell envisaged would be no more than a sheet mill but which Wick persuaded him should be made a complete puddled iron and finishing plant. Incorporated in November 1900 as the Youngstown Iron Sheet and Tube Company, in 1901–1902 it built 15 double puddling furnaces, a muck mill, skelp mill, three tube mills and six sheet mills on a 300-acre East Youngstown site. By 1904, output of sheets, skelp and pipe reached 1,000 tons a day. The next year, having recognized that the future lay in steel and

Table 4.4. Distribution of new sheet works built, 1900–1920

State or area	*Number*	*Annual capacity (thousand tons)*
East	2	250
Pennsylvania	9	643
Ohio	28	1,838
West Virginia	6	352
Kentucky	1	n.a
Indiana	4	597
Wisconsin	1	85
Total	51	3,791

Based on AISI Works Directories

not in wrought iron, the company removed "Iron" from its title and installed Bessemer converters and more rolling mill capacity, partly to supply its own requirements and partly to sell billets, sheet and tin bars on the open market. Before 1911 three blast furnaces and eight more sheet mills had been added.[15]

In 1912, another group was formed in the same area as an amalgamation of enterprises in complementary lines of business. The Brier Hill Iron and Coal Company had made iron for fifty years before it amalgamated with two Niles sheet mill companies. An open-hearth steelworks was built near existing blast furnaces to make 1,000 tons steel daily to be supplied both to the Niles mills and to independent rolling mills as billets, sheet and tin bars. In 1923, Brier Hill merged into Youngstown Sheet and Tube. In Canton the Berger Manufacturing Company was established in 1889. Eleven years later the Stark Rolling Mill Company was organized to supply it with steel sheets and going back a further stage in the manufacturing process, three years later a United Steel Company was set up to supply Stark with sheet bar. By 1916 the sheet and sheet bar companies had merged with the Canton Sheet Steel Company and another local firm to form the United Alloy Steel Corporation. Over a shorter period, the Massillon Rolling Mill Company, National Pressed Steel Company and Central Steel Corporation were formed and, by 1921, had

Figure 4.1. Hand-operated tin mill
Courtesy of Joseph Malborn

Figure 4.2. Old-style sheet mill
Courtesy of Joseph Malborn

merged. By 1922, these two groups controlled 273,000 tons of sheet capacity, much of it of high grade. Four years later, United Alloy and Central Steel amalgamated as Central Alloy, and by 1928 they became a major part of a new Republic Steel Corporation. In yet another case of enterprise by displaced managers, in 1919 H. A. Roemer, formerly with Canton Sheet Steel, headed a new sheet concern in Canton, the Superior Sheet Steel Company, which set out to build sheet mills capable of 50,000 tons a year.[16]

As early as 1878, the new Granite Iron Rolling Mills built sheet mills in the St. Louis area to use imported bar and operate with mill crews brought in from Wales. In 1895 a new site was developed at what later became Granite City on the Illinois bank of the Mississippi. Two open-hearth furnaces were followed by expansion involving more steel furnaces, sheet mills and galvanizing and tinning plant.[17]

Along the Ohio in 1889, George Verity, at 24 general manager of a grocery store, took over the Sagendorf Iron Roofing and Corrugating Company of Cincinnati, which was then in receivership. Two years later he reorganized it as the American Steel Roofing Company, renaming it once more in 1899, this time as the American Rolling Mill Company. In February 1901, this company made its first steel sheet at Middletown, in the Miami Valley 40 miles from the Ohio. During 1905 it acquired the 40,000-sheet capacity of a works at Zanesville in the interior of the state, built four years before. At Middletown it began to integrate backward in 1911 with an open-hearth plant, and turned out its first auto body sheet. It had already recruited two men who would be vital to its long-term success. Born in Cincinnati in 1880, Charles R. Hook started as an office boy with the Cincinnati Rolling Mill and Tinplate Company in 1898. A year later, this was absorbed into the American Tin Plate Company and Hook took employment as a mill worker in the Morewood works of American Tin Plate in Gas City, Indiana. He eventually married one of Verity's daughters and would be closely involved with this company for the next 58 years. In summer 1904, a man from a very different background asked for a job. Born in Middletown, and five years older than Hook, John Butler Tytus, son of a local papermaker, eventually played a key part in sheet mill technology.

Another instance of the migration of talent and enterprise at this time of rapid growth and reorganization involved W. H. Donner. Originally a flour miller in Indiana, he became associated with one of the largest tinplate works there, before moving to Pennsylvania to find a location for a new works which had a good fuel supply. Meeting the president of the Lake Erie Railroad, he agreed to buy a site for his plant on the Monongahela. The land company which owned it provided 20 acres and a cash subsidy. In spring 1897, construction of the Monessen works began; operations started a year later. Donner is said to have had vague ideas of a continuous process for his 8-mill plant, but in 1898 he sold out to American Tin Plate. Early in the new century he promoted another new works, at what became Donora.[18]

Along the 30-mile stretch of the Ohio above Cincinnati, two other works became important and featured in mergers which created significant players in the sheet industry. Located where the Scioto River joins the Ohio, Portsmouth had been an iron manufacturing center for 70 years before it was taken over in 1902 by two Wheeling firms, the Whitaker Iron Works and the Laughlin Nail Works, and began to ship sheet bar up the river to supply their finishing operations. Soon however, it was found that Pittsburgh was a cheaper source for semifinished steel, and it was decided that Portsmouth should become more self-contained. In 1911, it was equipped with sheet and jobbing mills and galvanizing plant. In 1920, Portsmouth, Whitaker-Glessner, the Wheeling Steel and Iron Company and La Belle Iron Works merged as the Wheeling Steel Corporation.[19] A few miles above Portsmouth, the Ashland Iron and Mining Company installed sheet mills at the small Kentucky town of that name in 1900.

A complex development process began with a 1893 merger, that of two Bridgeport, Ohio, firms, the Aetna Iron and Steel Company and the Standard

Iron Company. Among the chief officers of the first were John Topping and Isaac M. Scott, and of the second William T. Graham. After the formation of Aetna-Standard, operations in tinplate and sheet mills were continued until the company sold its tinplate works, sheet mills and bar mills, respectively, to the Tin Plate, Sheet Steel and American Steel Hoop combinations. After occupying leading posts in these companies, including the presidency of American Sheet and Tin Plate, Graham was chosen to head the newly formed American Can Company. Topping went through the same groups, moved for a time to the Tennessee Coal, Iron and Railroad and in 1908 became president of Republic Iron and Steel. Scott went to American Sheet Steel, but, having kept out of the tinplate industry for 15 years as required by the agreement which he signed when he sold his interest in the Beaver Tin Plate company of Lisbon, Ohio, to American Tin Plate in 1898, he became free in 1913 to be part of a new venture at Yorkville, that of the Wheeling Sheet and Tin Plate Company. In 1915 this merged into the Wheeling Steel and Iron Company. Five years later Scott was involved in the creation of the Wheeling Steel Corporation of which he became president.

Above Wheeling, another sheet mill center of great future significance was set up in this period. After employment in the Braddock Wire and then the Oliver Wire companies, in 1901 Ernest T. Weir entered the tinplate business, first as manager for American Tin Plate at its Monongahela mill and from 1903 at Monessen. By 1905 the eight mills of the Jackson Sheet and Tin Plate Company of Clarksburg, an isolated plant much farther up the Monongahela, had been idle for three years. They were offered to the United States Steel Corporation, but on 19 December its finance committee resolved that the proposal to buy them for $218,000 should be laid on the table. Instead Clarksburg was acquired in a private venture by Weir and his partner J. R. Phillips. After initial hardship, this new company made money, and Weir began to look for a better location for tinplate manufacture. After considering Pittsburgh and situations as far distant as Buffalo, he found what he was looking for in the valley of a small tributary of the Ohio in the panhandle of West Virginia. The tract was purchased in April 1909, and by the end of the year 10 mills were at work. The number was doubled in 1910; four more mills were added in 1914 and two in 1915. A cold strip mill was installed in 1913. The company added open-hearth furnaces, and in 1919 a blast furnace. By 1920, it had seven open hearths, Lake Superior ore properties and coal in Pennsylvania.[20]

The final major new center of Ohio Valley sheet capacity was the new, integrated works built by Jones and Laughlin at Aliquippa between 1907 and 1909. In the latter year it installed a large black plate plant to supply its tinning operations. By the end of World War I, sheet capacity was 130,000 tons.

The significance of the seven sheet operations on or near the Ohio and from Granite City to Aliquippa was to be considerable. Each was by 1920 a fully integrated iron and steelworks. Yet it had long been recognized that a river location, though better than a fully landlocked one, was inferior, and increasingly disadvantaged for iron- and steelmaking, as compared with the shores of the Great Lakes. Works in the interior would survive, expand and introduce major improvements in sheet mill technology in the course of the next few decades. Some would be eminently successful. But they represented very much a second best in terms of the distribution of the industry, and therefore of its potential profitability. In effect, they were responses to the legacy of the nation's already long and complicated industrial history rather than ideal choices for future manufacturing.

A striking instance of the coming together of existing plant, individual enterprise and an urge for expansion—all of which at a later date would help to create problems of a less than rational lay-out for the industry—was provided by the growth of the sheet and steel industry in Mansfield, Ohio. Mansfield was a city of less than 18,000 in 1900. At the heart of an agricultural district, it had a variety of industries, including farm machinery and foundry and machinery products, but no prominence in flat-rolled iron or steel. By 1909, the National Sheet Steel Company was building a plant there for 1,200 tons of high-grade sheet steel a month, largely for use in the electricity industry. In other words, it produced what was by no means ordinary sheet. By 1914 the Mansfield mill was operated by the National Rolling Mill Company, but in that year it was acquired by a member of the Davey family. After working in the tinplate industry of the Forest of Dean in Gloucestershire, England, the first member of the Davey family had come to the United States in 1888 and had been involved in four tinplate or sheet companies in the years since then. Six months after he arrived in the United States, his seven sons had followed him. For six years they worked in the Demmler Brothers sheet mills in McKeesport and then at the Niles works of the Falcon Tin Plate and Sheet company. Now, after acquiring Mansfield, the sons set out to make major extensions; being practical sheet mill men, they were successful from the start. In 1917 work began on a new sheet unit which came into production during Fall 1919, 12 mills producing a highly polished sheet. Finding difficulty in getting deliveries of sheet bar, they decided on backward integration, installing four 85-ton open-hearth furnaces, a roughing mill and 24-inch bar mill, specially adapted to produce 120,000 tons of high-quality material for their special sheet trade. The works was reckoned well-placed to tap the steel scrap produced over a considerable area.[21]

More important in the long-term development of the thin flat-rolled products industry was another firm created as a result of a complex of takeovers and new ventures by established producers. The 1912 formation of the Brier Hill Steel Company included the Empire Iron and Steel Company, a sheet-making concern. Within a few months, Empire's former president had organized a company to build a 35,000-ton tinplate works at Warren in the upper Mahoning Valley. Known as the Trumbull works, it grew rapidly and by 1920 had an open-hearth steel capacity of 400,000 tons. In 1917, some of those interested in Trumbull Steel established a $600,000 new sheet company, Liberty Steel, to build a 30,000-ton-per-year sheet plant in Warren to concentrate on high-finish sheet for the automobile and metal furniture industries. Two years later Trumbull merged with Liberty Steel, and this sparked off another venture. The former president of Liberty became head of the $3 million, newly formed Newton Steel Company which built sheet works at Newton Falls 18 miles west of Youngstown.[22] Trumbull and Newton would play important parts in the sheet trade of the 1920s.

Early in 1920 *Iron Age* surveyed the progress made by the sheet and tinplate trades since the early years of the century. According to its records, at the beginning of 1904 there had been 300 complete sheet mills and 43 more under construction; in tinplate, there were 335 mills with another 63 building. By 1920, the industry comprised 494 sheet and "jobbing" mills (the latter rolling heavier gauges) and 545 tinplate mills. Production in 1906 was estimated at one million tons of sheet and half a million tons of tinplate; now capacity had reached 2.5 million tons of sheet and 1.8 million tons of tinplate.[23] From a few of the many but often modest new projects or programs of reorganization of the first twenty years of the century, there came a search for improved methods of production. Their efforts would eventually cause a revolution in sheet-making technology.

An important incentive to innovation was the labor-intensive nature of sheet mill processes. Personal skill and judgment remained vital factors in this section of the steel industry long after mechanization had caused a decline in dependence on those qualities in most parts of the bulk steel trades. Sheet mill labor received high rates of pay, and it was conscious of its rights. In turn, employers endeavored to keep their men under control and to squeeze as much output as possible from them. In 1909 information was received in Vandergrift that outside organizations, possibly to the extent of 2,000 persons, and accompanied by brass bands, were expected to arrive in the town on Labor Day and to parade, presumably in the interests of organized labor. The local mayor, J. H. Chambers, who was an employee of the American Sheet and Tin Plate Company, issued a proclamation forbidding parades and marches.[24] Ten years later, at the time of a major national steel strike, 185 Finns at Weirton works were accused of being communists. They were forced by local vigilante groups to kneel in the public square and kiss the American flag before being run out of town. Ernest T. Weir then issued a formal order that his plants should never again hire a Finn. When researching conditions at American Sheet and Tin Plate during the recession conditions of 1914, the noted writer Ida M. Tarbell received a letter from an employee of that company. Although ill constructed and unsigned, it gave an insight into the increasing pressure of work in the sheet mills: "I am a worker of the US Steel Corporation of this place, am working 8 hours when I work. But mill has been off for six months and on one quarter to one third time for six months before that. We made 150 pair of sheet iron five years ago for day's work, but now if you don't make 300 pair they want to know what you have been doing and that if you can't do better than that they will hope to get some one that will. 12 years ago before the trust brought up the mills a good eight hour day was 105 pair of sheet iron they raised to 135 pair and then 150 pair and then to 165, but now if you can't get 300 they mark *rotten* or very poor after your day's weights. Hoping to see the day that I can do a day's work and come home feeling half-way good, I am yours."[25]

NOTES

1. Jeans 1902, 305.
2. *Economist* 25 August 1906, 1408; *IA* 14 June 1906, editorial; J. Malborn to author 1966.
3. "The steel requirements of the automotive industry," *AISI Yearbook* 1922, 381.
4. W. B. Gillies, "Promoting Demand through improving quality," *AISI Yearbook* 1938, 230, 231.
5. *IA* 22 September 1927, 779, 780; E. S. Lawrence, 211, 212; J. Malborn to author 1966.
6. Moody, *Analysis of Investments* 1915.
7. US Bureau of Census, 13th Census Vol IX, *Manufactures; Abstract of Census of Manufactures,* 1913.
8. Carnegie Steel Company Directors mins, 31 October 1912.
9. USS annual reports.
10. USS finance com. mins, 7 Feb 1905; *Engineering and Mining Journal*, 11 January 1908, 126; J. A. Topping to Dissolution Suit, 26 January 1914, 567; *IA* 18 January 1912, 213.
11. *IA* 1 January 1903, 39; *BFSP* September 1928, 1200; *Fortune* October 1938.
12. *ITR* 5 January 1911.
13. *IA* 4 April 1912, 892.
14. *ITR* 21 March 1912, 669; *BAISA* 1 June 1912.
15. *Engineering* (UK) Report of Philadelphia correspondent 13 September 1901; Jeans 1902, 159; *IA* 21 September 1911, 630; *Youngstown Bulletin*, November 1965, 5; YS&T in *20th Century Encyclopaedia.*

16. *BFSP* December 1919, 612.
17. *BFSP* January 1950, 83.
18. *Valley Independent,* 12 February 1972.
19. Rowe passim; May passim.
20. USS finance com. mins, 19 December 1905; J. F. Wall on E. Weir in *Dictionary of American Biograph; Supplement* 6.
21. *ITR* 3 January 1918, 52ff; *BFSP* January 1922, 71.
22. *IA* 25 January 1912, 259, 2 May 1912, 1089; 19 April 1917, 965; *Works Directory* 1920.
23. *IA* 1 January 1920, 50.
24. A. H. Beale, Vandergrift district manager to E. W. Pargny, president of AS&T 5 April 1915. *USS.*
25. Letter to E. W. Pargny 28 January 1915 in Tarbell correspondence 1912–1937. *USS.*

Chapter Five

A Great Divide

The Sheet Trade in the 1920s

Most commercially successful major changes in technology have been a response to important alterations in national economic circumstances. In the steel industry these have usually been shifts in demand, as with the vital impact of the spread of the rail network, or later of the huge construction programs for bridges or multistory buildings and the growth of petroleum extraction, transportation and refining. As illustrated by the technological changes with which these major changes in consumption were associated, innovations have widely ramifying further effects. As an industry adjusts to fundamental shifts in its markets, some existing companies will do so more successfully than others, new firms will emerge, grow and sometimes flourish, and older producers making an inadequate response or none at all, will stagnate, shrink or even disappear.

The 1920s saw the beginnings of another episode of this kind for the iron and steel industry. The reason for it lay in a fundamental shift in the pattern of demand for steel products, which marked a new stage in national economic development. Although there had been many earlier intimations of change, and the process was far from complete when the decade ended in the onset of the greatest economic setback in the nation's history, this decade was, as Rostow put it, "the first full experience of a society in the stage of high mass-consumption."[1] The most impressive indication of the scale of the divide over which America passed in the course of a few years is the total number of private motor vehicles registered in 1920 and 1929—8.1 million and 23.1 million respectively. For the steel industry, this marked the beginning of a new industrial revolution.

Overall the decade was one of high and rapidly rising prosperity. Truly nothing had ever succeeded like America, and its prosperity had never advanced so rapidly. The progress was marked in most parts of the economy which were major consumers of steel, but the gain varied widely. The value of new private and public construction all but doubled over those nine years from $10.5 to $20.8 trillion. At the other extreme, after astounding wartime achievements, shipbuilding collapsed, the gross tonnage of seagoing vessels launched from United States yards falling from 2.3 million gross tons in 1920 to 101 thousand tons nine years later. The experience of the railroads, so long a bellwether of the steel economy, was better, but scarcely inspiring. Rail freight revenues increased from $4.4 trillion to $4.9 trillion, but revenue from passengers fell from $1.3 to $0.8 trillion.[2] In 1920 railroads consumed an estimated 8.3 million net tons of steel, and the automobile industry 3.6 million tons: nine years later the respective figures were 7.2 and 7.6 million tons. Although the impact was much smaller, other consumers of thin flat-rolled steels were also mushrooming. It has been estimated that between 1910 and 1920 the output of refrigerators was no more than about 10,000—in total—and most of these involved wooden cabinets. Over the next four years the tally was 75,000 units. In 1925 came a transition to use of sheet steel, and refrigerator output in that year reached 100,000. In 1930, production reached a million units. By then this industry annually required 150,000 tons of sheet steel.[3]

Contrasts in growth had a direct impact on steel. Output of rolled iron and steel products in 1920 was 36.2 million net tons, by far the largest ever in a peacetime year; 1929 set a new record at 46.0 million tons. Between these two highs, most finished products showed an increase, but some did much better than others. Expansion in flat rolled products was outstanding. It was estimated that in 1929 81 percent of practical capacity was utilized in the rolled steel industry. For sheet and light plate the utilization rate was 89.2 percent. Yet, according to one computation, capacity to roll these products had increased by more than 49 percent since 1922.[4] Much expansion came from existing sheet works, and was obtained by adding new mills or investing to increase the output from existing mills, but there was also a large construction program for new sheet mills. In this period production by conventional technology was supplemented for the first time by output from continuous mills.

After World War I, the rapid increase in the number of new sheet mills suffered a setback in 1921 but in the next year light sheet production reached new high levels, and the expansion of plants continued. By the end of 1922, there were 659 sheet and jobbing sheet mills, with an installed annual capacity of 4.56 million tons. There was no net increase in mills in 1923, but after that each year numerous companies built extensions, so that by the end of 1926 there were 728 mills and a capacity of 5.32 million tons. Shipments went up even more—from 3.30 to 4.70 million tons. By the end of the following year, the total number of mills was 731, of which 682 were "sheet

Table 5.1. Production of rolled iron and steel and of individual products, 1920 and 1929 (million net tons)

	Total	*Rails*	*Heavy structural*	*Plate*	*Sheet*	*Strip*
1920	36.23	2.92	3.07	5.33	3.23	0.66
1929	46.00	3.05	4.54	5.62	5.89	2.80
Increase %	26.9	4.5	48.0	5.6	82.1	327.3

Based on AISI statistics

mills" producing lighter sheets and with an output averaging 8.67 net tons per 8-hour shift and 49 were "jobbing mills" rolling heavier gauges and averaging 22.56 tons per shift.[5] The industry was highly competitive. In the middle of the decade prices were sometimes depressed because so many producers of common sheets had upgraded their product and now also rolled auto body sheet. Up to certain widths, black plate mills were capable of upsetting the market in sheets and in times of slack demand plate mills cut into the markets for blue annealed sheets—as bar mills did into the field of narrow strip.[6] By 1929, at 6.4 million tons, the national capacity to roll sheets and light plates was one ninth that for all the main finished products. If the black plate used in tinning is included, the share of this sector in the industry's total capacity rose to 15.7 percent, far in excess of any other category. Yet in many ways sheet manufacture remained a highly atypical part of the steel industry.

In the first place, its operations were divided between a large number of producers, and the major iron and steel companies generally played a relatively small part. By 1927, there were about 40 firms in the trade, and the five main iron and steel companies controlled only 41.4 percent of the capacity in sheet and black plate, as compared with 86.4 percent in structural shapes, 82.2 percent in rails and 73.9 percent for plate. The one major exception was the American Sheet and Tin Plate Company. Through it, the United States Steel Corporation still accounted for about 25 percent of sheet capacity. In tinplate, it was even more important, having 256 mills compared with some 331 owned by independents.[7]

The reason for the large number of companies and the generally smaller participation of the majors was the same—even now, when it had become an important part of the whole steel industry, sheet manufacture was in general not technically integrated with iron- and steelmaking. A few firms owned open-hearth melting shops, but most of them continued to buy the bars they rolled down into sheet from bigger, integrated steel companies. Secondly, and largely because of its nonintegrated nature, a sheet mill could be relatively cheaply built. Yet it produced a relatively high-priced product. The apparent mismatch between these two characteristics was caused by the labor-intensive processes of manufacture still followed long after most branches of iron and steel manufacture had been mechanized. A leading trade journal summed up the labor situation in summer 1927: "The ordinary sheet mill is probably one of the largest employers of labor in the entire industry, and, what is more, the labor used in the sheet mill is of a highly skilled, high-priced kind, with an organization unusually well able to enforce its demands."[8] A few years before, substance had been put into those general statements. In testimony to the enquiries of the Senate Committee on Finance in connection with the Fordney Tariff bill, a representative of the sheet mills and associated jobbing mill firms maintained that directly or indirectly 80 percent of the cost of production of steel sheets was represented by the cost of labor. He reckoned that in 1920 the 42,000 mill men employed in rolling sheet bars down into black and galvanized sheet earned on average $1,910 yearly. (That year the average annual earnings of a manufacturing worker were $1,358.)[9]

Partly because of high labor costs, as well as more exacting uses, there was movement to higher-grade sheets. There was no very pronounced trend to increased output per mill. Yet sheet was a product for which demand was increasing rapidly, and, given the nature of its outlets, seemed likely to continue to do so. This—and the much divided ownership in sheet—reinforced a long-established tendency to expand capacity without close attention to a matching capacity to consume. Throughout much of the decade, the potential ill effects of this policy were hidden by even more rapid economic growth—between 1922 and 1926, for instance, sheet mill capacity increased almost 17 percent, but sheet shipments rose 42 percent—but even so there were occasional periods in which the problem became acute. As *Iron Age* rather laboriously put it in 1925, "Many persons feel, and not without some sound arguments to justify their reasoning, that the industry has expanded beyond present consumption demand and that it is only in what might be called abnormal periods of consumption that the industry can reasonably expect to operate at full capacity." Nonetheless, with increasing demand for sheet, a general expansiveness of the economy in the air, and many able, ambitious managers of existing mills eager to launch out on their own account, there was a built-in tendency to expand. Much later a general superintendent from the Sharon Steel Hoop Company recalled this particular form of upward pressure. It was a period "especially famous for putting down more sheet and tinplate mills, and renovating the older plants put down in the late 19th century and the early part of the 20th century. There was a crazy rush on the part of former mill managers to set out for themselves." For a time indeed "the money and power hungry former mill managers still felt they could out do the new-fangled hot strip mills." But in 1928, when the same man visited the Butler strip mill,

"I was convinced absolutely that our old-type sheet mills and tinplate mills were doomed to extinction in a very few years, and I reported my views to our president and other executives."[10]

During the 1920s some of the "old-style" sheet mills were applying a certain amount of technological innovation, in an effort to upgrade their product in order to command a larger share of the market. In 1922, Inland Steel pioneered the introduction of electric roll heaters for sheet and tin rolls, and in the same year mechanical doubling was first used in American black plate operations.[11] By 1925, some years after first reorganizing its Haselton mills in Youngstown for more efficient operation, the Sharon Steel Hoop Company virtually rebuilt the sheet mills there and installed another two mills. Old steam drives were replaced with electric motors. This company put in the first two-stage normalizing furnace. Annual capacity was increased from 70,000 to 120,000 tons. Most remarkable of all, this reconstruction program was carried through without ever shutting down the mill.[12] In January 1927, Inland brought into production a new, continuous sheet bar mill, and later that year the Newton Steel Company decided to build two continuous sheet normalizing furnaces each to deal with 160 tons of fully finished sheets a day. They were of the same general design as units installed by the Bethlehem Steel Corporation. After starting four new mills, Bethlehem built a continuous sheet bar and skelp mill at Sparrows Point.[13]

In a rapidly expanding industry of many small producers, operating in a generally growing but variable market, price stability was difficult to secure. At the beginning of the decade, paternal oversight by the United States Steel Corporation, still operating under the chairmanship of Elbert H. Gary, provided some discipline, at least for the bigger firms. This situation, and some of the benefits it brought, was highlighted by James Campbell of Youngstown Sheet and Tube at the annual meeting of the American Iron and Steel Institute in May 1921: "Somebody else fixes the price on pipe and sheet and our other products. The Judge does it to a large degree, and I am perfectly willing that he should do it; because he never makes prices below a point at

Table 5.2. The sheet mill industry, 1922–1926 (quantities in thousand tons)

	No. of hot mills	*Capacity*	*Shipments*	*Utilization rate %*
1922	659	4,560	3,300	73
1923	666	4,730	3,987	84
1924	702	4,900	3,639	74
1925	716	5,250	4,452	85
1926	728	5,325	4,700	88

Based on *IA* 11 April 1927, p. 1092

which we all can live. (!)"[14] By that time Campbell was in his late sixties, and his contentment with the situation was unlikely to be shared by younger, expansion-minded sheet mill managers. US Steel-led price maintenance was breaking down anyway. In depression conditions early in the decade price-chiseling went on, and in 1924 came the abolition of the Pittsburgh Plus system which had helped the Corporation to control pricing in much of the industry. (American Sheet and Tin Plate had gone so far as to supply its rivals and the National Association of Sheet and Tin Plate Manufacturers with booklets containing elaborate tabulations of prices and freight rates for all locations.)[15] In summer 1927, Judge Gary died. His successors did not provide the same sort of price leadership.

In 1928, the engineering consultant Leon Cammens spelled out the price situation inherent in the structure of the sheet industry:

> In sheet rolling as constituted at present the costs for the same class of products are substantially the same in the smallest and the largest plant, while the investment required is so moderate that practically any experienced man with a little money can start making sheets. The small producers as a rule can work cheaper than the large ones, because the owner or the principal stockholders of a small mill are as a rule also its managers and consider the return on their investment largely in the way of salary, while the large mill is managed through hired executives and must make a salary return in addition to a return on stocks. Moreover, the small mill is hungry for business and often willing to work for a lower return. If we add to this that the small mill often does not know its own costs, it will become obvious why it can take away business from the big mill on a price basis and is apt to establish a general level of prices which would not be accepted in industries controlled by big units. It is because of this that except in a very few cases the sheet business has not contributed the share of profits that would be warranted by the volume of its output. In other words, the smaller sheet mills have been willing to do business on other bases than that of profits and the big mills had to follow the small mills because there was nothing else to do.

Ominously, he added, "It would appear from this that the economic salvation of the sheet industry lies in a complete reorganization of its method of production along such lines as have proved successful throughout the entire American industry."[16]

In the light of the above, it is understandable that there were problems time and again during the 1920s with sheet prices. For instance, in 1924 and the early months of 1925, the price of 22-gauge sheet at Pittsburgh fell by $19/ton. This was in part attributed to overproduction, as so many mills which formerly rolled common sheets had moved in the last year or two into auto-body grades.[17] 1926 was a record year, but 1927 shipments to the automobile industry fell by over 26 percent and at one time utilization of sheet mill capacity was reported as down to 62 percent.[18] By early that year unintegrated mills were suffering a cost/price squeeze, a sheet prices being depressed while they were still paying a high price for purchased sheet bars. Some firms were in financial trouble. Such circumstances sparked attempts to improve prospects by amalgamations. In spring 1927, two sheet firms established at the beginning of the decade, the Chapman-Price Company of Indianapolis and the Superior Sheet Steel Company of Canton, Ohio, having a combined capacity of 110,00 tons of sheets, merged with the Kokomo Steel and Wire Company of Indiana to form the Continental Steel Corporation with a capital of $15 million. Kokomo had 180,000 tons of open-hearth capacity, a disproportionately large blooming mill capable of 400,000 tons a year, but a finishing capacity of only 120,000 tons in the form of billets, rod and wire. It was decided to make large extensions to steel capacity there and to build a sheet bar mill to supply the Canton and Indianapolis mills.[19] Earlier in 1927, an even bigger scheme was taking shape for amalgamation and streamlining in Ohio sheet mills.

In January 1927, W. A. Thomas, former president of Brier Hill Steel, which in 1923 had been taken into Youngstown Sheet and Tube, announced that negotiations, backed by a banking group, were underway for a merger of nine sheet firms, some of which had been in difficulties during the preceding year. The nine companies owned 100 sheet mills, almost one in seven of the total in the trade. Five were in the Mahoning Valley between Niles and Lowellville. The companies varied considerably in size, and only two, Mansfield Sheet and Tinplate Company, and Sharon Steel Hoop Company, possessed steel capacity. They were able to supply their own mills, but to meet the needs of other firms in the new group it was proposed to buy sheet bars from YS&T, Republic or Corrigan-McKinney. Within a few weeks Sharon Steel Hoop and two of the sheet companies withdrew from negotiations, but the Empire Rolling Mill Company of Cleveland joined them. Annual sheet capacity in the seven companies which remained was 456,000 tons. Mansfield's 185,000 tons of open-hearth capacity and 165,000 tons of sheet bars were now the only supporting capacity within the group. The merger did not materialize, largely because of failure to agree on valuations, but months later efforts were made to revive it, and in January 1928, after more meetings and the sorting out of main points of disagreement, five companies from the original nine and the Empire concern were incorporated as the Empire Steel Corporation, under the presidency of W. H. Davey, head of Mansfield Sheet and Tinplate. Empire Steel owned 60 mills with an annual capacity of 400,000 tons, for which its Mansfield works was expected to supply about half the sheet bar, the remainder being bought in the open market. It was remarked that such amalgamations

Table 5.3. Proposed merger of Ohio sheet mill companies, January 1927

Company	*Location*	
Newton Steel*	Newton Falls	20
Thomas Sheet	Niles	12
Waddell Steel	Niles	8
Falcon Steel	Niles	8
Ashtabula Steel	Ashtabula	8
Mahoning Valley Sheet+	Niles	8
Sharon Steel Hoop	Lowellville	9
Mansfield Sheet & Tinplate	Mansfield	19
Superior Sheet Steel*	Canton	8

* Companies which quickly withdrew from merger discussions.
+ Withdrew at a date unknown
Source: IA 20 January 1927, p. 233

often disappointed hopes of cutting costs for their material needs might be purchased more cheaply from outsiders—in which case others and not the company had to bear the charges for underused or idle plants.[20]

In 1928, activity revived strongly and then reached new heights. A record was set in 1929 at almost 5.9 million net tons, 82 percent above that of 1920. Such conditions encouraged more plant extensions and the construction of new mills. For example, during spring and summer 1929, Continental Steel rushed through installation of six new sheet mills at Kokomo, and Apollo Steel announced it was considering building four to six open-hearth furnaces to back up its 125,000-ton mill capacity for black, galvanized and blue annealed sheets. In July a Tarentum Steel Corporation was formed with a capital of $3 million to build a new six-mill sheet works to produce auto grade sheets on the site of the former Pittsburgh Plate Glass Company, which had been abandoned 15 years earlier.[21] Yet, though 1929 was a year of record activity, threats to the longer-term prosperity of sheet mills formed a sobering backdrop. The characteristic response to good times of investing in additional capacity seemed to point to perennially uncertain futures, but there were other, bigger, problems which were quite beyond the control of individual companies. If a really sharp and sustained setback in the consumer boom occurred, or if capacity extensions in thin flat-rolled products were to be made not, as in hand mills, in many separate small increments, but rather in very large additions of high-quality, low-cost output, the traditional industry might soon be in serious trouble. The prospect of both eventualities occurring together was too painful to contemplate. During fall 1929, it became clear that this was happening. For the sheet mills, the 1920s had been a decade of success; as it ended, they entered a crisis of unimagined proportions.

NOTES

1. Rostow 1971, 219.
2. *Historical Statistics of the United States*: Lloyds Register.
3. *IA* 27 August 1931, 557.
4. *America's Capacity to Produce,* 263, 264, 3035.
5. *IA* 2 April 1924, 112, 113, 1 January 1925, 86, 14 April 1927, 1092, 29 December 1927, 1828.
6. *IA* 2 April 1925, 986, 13 October 1927, 1040.
7. Pittsburgh correspondent in *ICTR* (UK) 14 January 1927, 55, 29 April 192, 678; McCallum 1931, 131; *IA* 14 April 1927, 1092.
8. *BFSP* June 1927, 267 quoting *Mechanical Engineering.*
9. Quoted *ICTR* 7 October 1921, 502; JR Commons 1929.
10. *IA* 14 May 1925, J. Malborn to author.
11. Goodale, 1931, 294.
12. Recollections of J. Malborn as general superintendent of Haselton 1919–1931; *IA* 5 March 1925, 703.
13. *IA* 18 August 1927, 394, 15 December 1927, 1686; *BFSP* January 1927, 50, January 1928, 61.
14. *AISI Yearbook* 1921, 28.
15. G. W. Stocking, 1954, 52.
16. Cammens, 1928, 5.
17. *IA* 2 April 1925, 986.
18. *Steel* 16 April 1945, 82, 83; quoted *ICTR* 17 August 1928, 222.
19. *IA* 8 April 1929, 1236.
20. *IA* 20 January 1927, 233, 17 February 1927, 523, 3 March 1927, 676, 2 June 1927, 1640, 27 October 1927, 1177, 5 January 1928, 901, 1 March 1928; Frasier and Doriot 1932, 255.
21. *ITR,* 26 September 1929, 779; *IA* 18 April 1929, 114; *ITR* 18 July 1929, 174.

Chapter Six

The Search for a Practicable Continuous Sheet Mill

In late April 1925, the president of Carnegie Steel replied from Pittsburgh to an inquiry he had received from the manager of US Steel's Bureau of Safety in New York about inventions in the steel industry: "There have been no basic inventions since the time of Bessemer, in the Bessemer process, and Siemens-Martin in the Open Hearth process, but those that stand out in the way of cheapening of product and made mass production possible are: the Jones mixer for handling large quantities of pig iron in the molten state instead of casting the iron into pigs and re-melting; the Dinkey controller which made possible the use of electric motors in steel mills; the Wellman charging machine that lowered the cost and increased the product of the Open Hearth process; the Brown ore bridge that made it possible to handle economically the large quantities of ore necessary for modern blast furnaces, and along this line the Hulett unloading machines for taking ore out of the boats and the Hulett car dumper for unloading ore from cars quickly; the Uehling pig casting machine which did away with the large amount of labor of the old days when iron was cast in sand and chills in the cast house. . . . These are the only improvements and inventions that I can recall at this time."[1] It was a respectable list, but completely ignored a revolution which was already underway and would soon far outshine almost all of the later aids to mass production he had identified.

As seen above, for many years sheet production had been acknowledged as an anomalous section of the rolling mill industry, with methods of manufacture at odds with the American principle of employing as much machinery and as little labor as possible per unit of product. Consequently sheet mills had high costs and relatively low productivity compared with most other types of rolling mill. Other sections of the industry were concerned with larger—often much larger—outputs, and because of this some had used continuous mills for many years with great economy in effort and much-reduced labor costs. Three years before the 1925 list of inventions quoted above, a paper was given to the Engineers Society of Western Pennsylvania by J. W. Sheperdson, of the Morgan Construction Company of Worcester, Massachusetts. His firm, along with the English wire maker J. P. Bedson and the immigrant mill engineer William Garrett, had over many years transformed the manufacture of bars, rod and wire. Sheperdson told his engineering audience, "As to the future of the continuous mill, we know of no way to judge other than by the past. Whenever, in the past 40 years, the demand for any product has reached such proportions that a single organization could market the output of a continuous mill, a continuous mill for that product has made its appearance."[2] For thin flat-rolled steel, the huge increase in demand during the 1920s seemed to indicate that a propitious time for continuous mills had arrived. Indeed, by the time of Sheperdson's address the prototype mill was already at work.

There were a number of problems to be faced in any attempt to roll sheets continuously or even semicontinuously—the piece in the latter type of mill *not* being in two successive sets of rolls at the same time. Hand sheet mills made use of "packs," a number of sheets placed one on top of the other. The pack retained heat, and enabled thin sheets to be rolled. At the end of the process the pack was opened to be separated into the individual sheets. Early experiments in more continuous rolling also used the pack method, but found there were difficulties in opening the packs. However, there was insufficient power available for the rapid rolling necessary to retain heat in a single piece. At that time also the accuracy of roll contouring could not be relied on.[3] Such problems had not prevented a number of earlier efforts in the direction of more continuous rolling. Paradoxically, though they were eventually to be successful, the search proved something of a dead end. The way of advance would lead in a different direction.

Douglas Fisher suggested that an unsuccessful effort to roll sheet in something like a continuous mill was made by H. F. Mann in Pittsburgh as early as 1865. Ten years later another attempt was made by S. R. Wilmot in Bridgeport, Connecticut. There was a much bigger, longer-lasting and more promising effort in central Europe in the 1890s. It was defeated in part by its unfavorable commercial setting. From July 1892 continuous rolling operations were begun in sheet mill finishing stands, placed nine feet apart, and added to a plate mill at Teplitz, Bohemia. The works was owned by Karl Wittgenstein, a major Austrian steelmaker and father of the future philosopher, who had been born three years before. In the Teplitz experiment it was necessary to reheat the steel during the process, but it was never allowed to cool completely between the ingot and the finished sheet. By November 1892, 30 tons of sheets were being made in 24 hours, but this was not a marked advance on what was possible in three shifts on a conventional mill. In its early days at least, the Teplitz mill could only roll thicker sheet, ranging down from a little under 2 mm to about 1.6 mm. There was one man on the finishing stands, who had to adjust the rolls as required, but there were also men who had moved on from the roughing trains.[4] Later reports suggest that sheets in lengths of up to 60 feet and as wide as 50 inches were rolled. When the professor of commercial engineering from the Carnegie Institute of Technology, S. B. Ely, visited Teplitz—he later reckoned it was in 1910 or earlier but in choosing this date seems to have underestimated how long ago it had really been—he was impressed to find that the piece being rolled could be passing through five stands at once and not tear or crumple. In the five stands of two high rolls it was reduced in thickness by 58 percent from 7mm to 3mm. Then, by passing the piece through two 3-high finishing stands, the mill rolled 12-gauge sheet.[5] One serious practical problem was variability of gauge. The power of the steam engines used to drive the mill proved inadequate. Perhaps a greater unfavorable factor was the operating context—the scarcely dynamic economy of the Austro-Hungarian Empire—for demand was insufficient to enable the mill to run steadily on a few gauges. Commercially speaking, the process failed, and it was abandoned in 1907. Initiative now passed to an industry in which the operating context was much more favorable.[6]

Consumption of sheet in the United States was already vastly bigger than that in the economies of Europe, and was increasing rapidly. Advances in electrically powered mill drives removed another of the problems which had plagued Teplitz. From the early years of the new century, purposeful moves were made towards more continuous operations. Late in 1901, there were press reports of the introduction by a Pittsburgh firm of a new sheet rolling process said to cut costs by $5–$8/ton—at a time when black sheets were in the upper $60s or low $70s a ton. Nothing permanent came of this particular "breakthrough," but a few months later the same journal announced there was little doubt that the day of "revolution" for sheet was at hand. It reported that US Steel had authorized a six-figure expenditure on the commercial trial of a new process. A successful outcome would have far-reaching consequences. "Should the new process be a success . . . the changes which would follow would affect not only the technical side of tinplate and sheet manufacture, but would involve labor questions and the relations of the sheet and tinplate companies [of US Steel] to their new and now numerous competitors as well." The new process would be controlled by the corporation and would involve a saving in cost of manufacture so great as to practically eliminate competition. "These are possibilities of the near future, and in any event it is certain that the long-delayed improvement in methods of rolling sheet and tin plates cannot much longer be put off."[7] The effort referred to marked the beginning of trials which lasted over 16 years. The gain was less than had been anticipated by this fulsome journalist.

Four conditions influenced this US Steel effort for a radical improvement in sheet production. First, sheet mill operations were recognized as cost intensive by a top management, which, having largely been brought over from Carnegie Steel, was committed to a relentless search for greater efficiency through mechanization of processes. Second, the inclusion in US Steel of American Sheet Steel and of American Tin Plate gave it not only a preeminent position in both trades but also a great deal of excess capacity. Rather than scrap all the surplus plant capacity, why not use some of it to try to improve production methods in this important but, technologically speaking, lagging sector of the industry? Third, a more immediate incentive to try to mechanize processes was provided by the leading role played in the labor disputes of summer 1901 by those employed in this part of the Corporation's operations. Finally, along with other works acquired in spring 1901, this sector brought in the expertise of a large number of formerly separate mill managers.

In 1894 when the new Beaver Tinplate Company began constructing a works at Lisbon, Ohio, the president of its small operation was Charles W. Bray. By the late 1890s, Beaver had seven hot and seven cold mills and tinning facilities. It was included in the American Tin Plate Company in December 1898. At that time, Bray was not made a director or leading officer of AT, but under US Steel he became chief engineer. The parent company provided him with incentives to try to improve mill performances. In April 1902, the Finance Committee approved a recommendation that his annual salary be increased to $15,000.[8] This occurred at an early stage of efforts under his direction and extending over a number of years to improve rolling of sheet and black plate. The first attempt to develop new methods was made at American Tin Plate's Monongahela works, in Pittsburgh. Construction of a new mill, designed in 1901, was undertaken the following year. It began experimental rolling in October. The aim was to introduce continuous methods into the roughing stage of tinplate manufacture, starting with standard-size tinplate bars but breaking them down into tinplate stock mechanically. To this end, Bray arranged four continuous bar heating furnaces to feed a roll train of eight two-high black sheet mills arranged in tandem. After passing through these mills, the "pack" was mechanically doubled and then finished in ordinary two-high mills. In the January 1903 review of the sheet and tinplate industry, the Pittsburgh correspondent of *Iron Age* described the black plate operation as a success. Saving in labor costs as compared with normal practice was put at about $2/ton. In May 1901, the average price of black sheets was $71.68 and of tinplates $93.86 per gross ton. Comparatively speaking, therefore, the gain was small, although over the large tonnages of these products, it would be significant in total. But reduced labor costs were canceled out by losses due to an increased number of "seconds," extra scrap and roll breakages. After operating more or less continuously from January 1904 to July 1905, it was decided to try a different, more ambitious approach.

At the end of 1902, US Steel absorbed Union Steel and with it inherited machinery which, with additions, became the basis of the second attempt to improve rolling methods. The patents for the improvements were again those of Bray, now chief engineer and first vice president of American Sheet and Tin Plate. In November 1905, an experimental mill was put up at South Sharon (later re-named Farrell) in the Mercer mill, reckoned one of the finest works in AS&T. It used sheet bars instead of the slightly lighter tin bars, and employed nine rather than eight stands of rolls, but the process again ended with mechanical doubling and finishing operations on an ordinary two-high mill. Ely described the Mercer experiments as a "most thorough" trial but maintained the sheet was never in two stands at once—in short, the plant was not strictly speaking a continuous mill.[9] The Mercer mill operated intermittently from November 1905 to October 1907, was idle throughout 1908—presumably because of the severe recession—and was partially at work again in 1909 and intermittently until spring 1910. Problems were caused by a tendency for both bars and sheet "breakdowns" to stick as they passed through the mills. It is said that savings as compared with old hand mill operations were about $1.80/ton, but, as with the Monongahela experiment, the gain was more than counterbalanced by a number of extra costs—$2 for maintenance, $1 for rolls, $1 on account of a greater yield of scrap and $0.80 for various other items. Overall, therefore, costs in the Mercer trials were about $3/ton more than those in conventional practice. In May 1910, the experimental mills were closed, and during 1911, "the Bray semi-continuous mill" was dismantled and operations at Mercer returned to older methods, but with a concentration on what were called "specialties." A few months later, *Iron Age* seemed to draw a line under the search for continuous sheet mill operations when it concluded, "All the various efforts along this line have proved failures, and enough knowledge has been accumulated to suggest the probability that the substance of the methods of rolling sheets will not be changed."[10]

In fact, a final American Sheet and Tin Plate experiment was made in 1917. It seemed appropriate that it should take place at Vandergrift, an exceptional operation by virtue of its high reputation, large capacity and possession of an open-hearth plant. Such conditions might have been expected to lead on to success. The new procedures were given the name of the "Tipperary" system, presumably reflecting the wartime song—or perhaps also the long production sequence. This now involved linking three two-high mills. Useful experience was gained, but the output was inferior to that from some old-type mills, and trials went on for only a short time.[11] US Steel then gave up major experiments in sheet mill practice. The failure over the years of its three American Sheet and Tin Plate experiments was well known. Even so, it acted as a spur to others who would carry their experiments through to commercial success.

Looking back on this long period of experiment with continuous rolling, some commentators at least thought that the practical problems had been largely solved but that the work was frustrated by an unfavorable business environment. In mid-1927 *Blast Furnace and Steel Plant* reprinted an article on "The Continuous Rolling of Sheet Steel" from *Mechanical Engineering* which epitomized this attitude. It suggested that the AS&T continuous sheet mill "some years ago" seemed to have been a mechanical success but was a failure commercially, probably due to the fact that it had been built before the market was sufficiently large: "Continuous mills can be profitable only when a steady and substantial tonnage can be fed into the rolls day in and day out without much changing. A continuous mill is substantially a heavy tonnage producer, and will pay only if used as such." When the AS&T trials were made, orders had run to only a few carloads at a time.[12] This comment ignored both the many threads of evidence that there were also serious "mechanical" deficiencies in the AS&T experiences and the fact that growth in demand for sheet had already occurred on a large scale. Another consideration was that sheet mill capacity had increased more rapidly than had any prospect of reliable year-in, year-out demand. But, though this circumstance introduced more commercial uncertainty, it also provided encouragement for experimenters with methods which could increase labor productivity and help the successful company carve out a market for large tonnages of a higher-quality and more standardized sheet.

In the 1920s the continuing increase in demand for sheet steel and pressures for ever higher qualities in terms of uniformity of gauge, increased consistency of chemical and physical structure to enable ever more demanding pressings, as well as excellence in surface finish, encouraged further searches for a better way to make large tonnages more cheaply. Success promised increased profit to those companies which pioneered. The American industry was the only one in the world where growth in the sheet business was on anywhere near this scale. Between 1920 and 1929, British output of sheets less than 3 mm thick went up from 402,000 to 666,000 net tons. By 1929, German output of sheet of this gauge was 1,089,000 net tons. American production of black sheet over the same nine years increased from 3,232,000 to 5,886,000 net tons. In 1929, one account of the international situation observed that "at present, according to some important domestic interests, Europe's best sheet is no better than the single pickle grade made in this country."[13] It was an evaluation not accepted by all. But whatever the truth in this matter, given growth of consumption on such a scale, the introduction of a continuous mill to make sheet looked an increasingly attractive proposition. Reaching this goal required the solution of a number of serious technical problems. Naturally those who tried to tackle them were cautious both in publicizing their intent and in revealing the headway they were making.

In January 1920, *Iron Age* ran its usual review of the previous year's operations. Luty, who had written about Bray's experiments almost 20 years earlier, commented on the current state of the sheet and tinplate trades and saw no imminent prospect of a breakthrough to radically new methods of production: "For a long period of time, many men were seriously engaged with the problem of devising mechanical methods for rolling sheets and black plates for tinning. . . . At the present time, however, sheets and black plates are rolled in practically the same style as formerly. The steel is still handled by the tongs of workmen. The roller feeds the steel to the rolls, the 'behinder'—as they call him in Wales—catches the piece or pack and passes it back over the rolls to the roller, the matcher and doubler handles the material with his tongs, and the heater usually performs much the same work as formerly, though there have been some important improvements in the heating furnace. It may be said that the efforts have failed, that the work of a quarter century ago is still done by hand, and not mechanically." Yet the possibility of change was also in the air. In spring 1921 Sumner B. Ely of the Carnegie Institute of Technology gave a paper reported in the steel press as "Improved methods of Rolling Sheet Steel—Possibility of Use of Continuous Mills or Other Labor-saving Devices . . ." On the other hand, when in 1922 the editor of *Iron Age* reviewed the last 12 years of progress in iron and steel, he too, like the manager within US Steel quoted earlier, identified little novel or noteworthy in this section of the industry: "Sheet-mill construction changed little except that the electric drive became universal."[14] By the time these assessments were written, the drive toward revolutionary change in sheet was already far advanced.

At more or less the same time in 1901 when US Steel was taking shape with a raw steel capacity of almost 10 million tons, the diminutive American Rolling Mill Company was building its first works, having one steel furnace and an annual finished steel capacity of 25,000 tons. The capital of the Corporation was over $1 billion; that of American Rolling Mill $500,000. Yet two decades later it was this former industrial pygmy which pioneered the breakthrough in continuous sheet rolling. Key roles in this process were played by the plant superintendent of Middletown works, Charles Hook, and above all by the young university-trained man who had joined the company as a doubler in a sheet mill in 1904. John Tytus had advanced to become a regular rougher in the sheet mill crew. In 1905, he moved from Middletown to become assistant manager of the company's Muskingum works, but in 1910 he returned to the headquarters plant to install sheet mills in the Middletown East Works. Before coming into the steel industry, Tytus had worked for a time in his father's paper mill. At some indeterminate stage in his Armco career, he reportedly remarked to Hook how the sight of a Fourdrinier paper mill in operation had stimulated his ambitions for his new industry: "Some day, Charlie, we'll be making sheets in long strips like they make paper."[15]

Early in March 1915, Hook, as general superintendent, reported to Verity on a conference of leading officials of the company which had spent the day considering ideas for mill improvement proposed by Tytus. Hook thought the scheme looked "very practical" and reckoned it would save at least $2/ton as compared with costs in their 9-inch mill. Within a year as output built up the saving might be $5/ton. They would need to invest in a continuous furnace but had already done a good deal in this direction. The necessary screw downs and guides would have to be worked out in their own workshops. Hook estimated total research and development costs might be $3,500–$4,000, but he would know better after another six to eight weeks of work. He ended by saying, "We feel that this is one of the most important propositions that has been up for consideration in many years."[16]

During World War I, priority allocation of steel capacity for making shells gave Tytus an opportunity for further experiments with idle sheet mills. This led to the conclusion that careful coordination of contours in successive rolls in a continuous train would be an essential step towards success. He began work planning continuous-sheet production at Middletown, but the urgent need for maximum output of sheets to meet the demands of the postwar boom frustrated his hope of building an improved mill.

A new direction for advance opened up when, during 1920, George Verity, as chairman of the company, received a low-priced offer of the properties of the Ashland Iron and Mining Company of Kentucky. Ashland had local coal supplies, two blast furnaces with an annual capacity of 180,000 tons, a six-furnace open-hearth plant of 270,000 tons, a modern blooming mill but relatively small finishing capacity in the form of slab, billet, bar, and sheet mills. Late in 1920 the offer of Ashland was turned down, but a year later the now-depressed state of trade caused it to be renewed with even more attractive terms. This time Verity's board accepted and the deal was completed in December 1921. It was realized that the imbalances in Ashland capacity gave scope for experiments with the new style of mill which Tytus contemplated, and by summer 1922 Armco was said to be contemplating spending $6 million or so on finishing capacity there. In fact, it authorized expenditure of $10 million to build a continuous sheet mill, thereby, as Verity later admitted, staking the whole future of the company on the success of a risky venture. After carrying out more design work at Middletown, Tytus and his team of engineers, amounting to some one hundred, moved to Ashland. There, in strict seclusion from public view, they struggled with a host of problems common to all major projects, as well as many unique to their own. Gradually they made headway.

Iron Age on 3 January 1924, in its usual long annual review of the state and progress of the iron and steel industry, included a brief sentence: "The American Rolling Mill Company, Middletown, Ohio states that extensive additions have been made to its plants at Ashland, Kentucky, definite information concerning which cannot at present be given out."[17] Three days before that statement appeared, on the last day of 1923, a first major new installation had been completed at Ashland. This heavy sheet mill was followed over the next month by the remaining units. The manufacturing process developed by Tytus and his engineers involved four main stages: the production of slabs in the existing blooming mill; rolling the blooms down into what they called "bar-plate" (a wider, thinner piece than the sheet bar used in conventional sheet mills); followed by a seven-stand roughing sheet mill and ending with a five-stand and essentially continuous two-high sheet finishing mill.

There were many teething troubles, including broken rolls and housings and the buckling of sheets as they passed through the stands, but, after only a month, the new mill was rolling 350 tons of sheet a day—when an average hand mill turned out 8.5 tons in an eight-hour shift and even the large, so-called "jobbing mills," with their labor-saving devices, no more than about 22.5 tons a shift. Within three years daily sheet production had reached 1,400 tons.

Compared with the hand mills, the Ashland continuous mill achieved economy of both labor and power and reduced the loss which was involved in hand mill operations in both scrap and second-grade material. Product quality was good so that the old mismatch between the attainment of quantity and of quality began to disappear. Indeed one authority quickly recognized that "quality is actually dependent on tonnage in that uniformity of gage and surface call for continuity of operation."

Tytus had solved one of the basic problems in rolling a thin, wide product such as sheet. Earlier experiments from Teplitz to Mercer had not faced up to the problem of roll contouring, and consequently the surface shape of their breakdowns had not readily fitted the contour of the finishing mills through which they had to pass. It was not easy to make the contours in successive stands of finishing mills compatible, but by paying attention to the difficult point at which a piece was in successive rolls at the same time, Tytus had realized that earlier innovators had been wrong to strive to produce rolls which

Figure 6.1. An artist's impression of the Ashland continuous sheet mill in 1923
Courtesy of AIST

were perfect cylinders. This might work with narrow sheets, but such rolls could not produce good wide, thin sheets. If the process started with a flat piece, the sheet would move to the side of the rolls and the product would be unsatisfactory, but if a rolling sequence began with a piece slightly convex in cross section, it would pass in a straight line through the mill, each pass slightly reducing its convexity so that at the run-out end of the mill the sheet would be flat.

It was indisputable that Tytus's achievement at Ashland was outstanding. An *Iron Age* editorial observed it had been "much talked of but little known," hailing it as "epochal" and "a monumental example of the scientific approach to a major manufacturing problem." Yet the Ashland project was by no means fully satisfactory. The mill rolled relatively short lengths and its operations were only partially continuous. In the bar-plate mill the piece was never in more than one stand at once, and this part of the mill sequence was located 228 feet from the sheet mill stands.[18] For a time, however, it seemed that the Tytus continuous sheet mill would be the way forward, and the American Rolling Mill Company acted to consolidate its position as the pioneer of a revolution. On 12 October 1926, it was granted patents covering the continuous sheet rolling processes. Next January, Verity said his company needed to "round out" both Ashland, where they were planning to spend $1.5 million on two new open-hearth furnaces, and the Middletown works. By March, the company was wholly in control of Ashland, having bought out the minority interests of T. M. Adam. Four months later, Verity announced planned expenditure of another $10 million over the next 15–18 months, which would include building a second Ashland mill. Immediately afterwards, it was revealed the company had placed orders with United Engineering and Foundry for a continuous mill at Middletown.[19]

In November 1926, recording the patenting of the Ashland processes, *Iron Age* once more acknowledged the significance of what had been achieved: "This is regarded as the first radical development that has been successful in an industry that is over 200 years old."[20] Yet despite its major technical and commercial significance, the mill developed by Tytus and his team of engineers was not to be the father of the wide continuous strip mill. The more direct ancestry of that process, as its name suggests, was to be found in advances in a mill rolling not sheet but steel strip. The Ashland continuous-sheet mill had far outperformed the hand mills, but it coexisted with them; the new continuous wide hot strip mill would soon to begin to replace them.

NOTES

1. H. D. Williams to C. L. Close 30 April 1925. *USS.*
2. *Mechanical Engineering,* quoted *BFSP* June 1927, 268; J. W. Sheperdson of Morgan Construction quoted *ICTR* 31 March 1922, 448.
3. *IA* 9 December 1926, Editorial "The Sheet-Strip Evolution."
4. There is a short contemporary account in the Foreign Abstracts of the Institute of Civil Engineers summarized in *BAISA* 19 July 1893.
5. SB Ely in *Proceedings of the Engineers' Society of Western Pennsylvania* quoted in *ICTR* 21 June 1922, 79.
6. Hogan, 845; Fisher 1963, 143; Ely quoted *ICTR* 1 September 1922, 297.
7. *ITR* 5 December 1901, 26, 27, 21 August 1902, 29, 30.
8. *BAISA* 10 April 1895, 85; USS Finance Committee mins 22 April 1902; *IA* 5 January 1903, 13 in BET Luty of Pittsburgh on sheet and tinplate—"Extension of the Bray process"; *IA* 7 Jan 1904.
9. Ely op cit 1922, 79.
10. *IA* 4 January 1912, 64, 8 August 1912, 311; S. Badlam on evolution of the hot strip mill to *AISI* 1927, 387–95 and in *IA* 3 November 1927.
11. J. Malborn, in correspondence.
12. *BF&SP* 1927, 266–69.
13. *ITR* 14 April 1929, 265; *IA* 16 May 1929, 1371, 1372.
14. BEV Luty in *IA* 4 January 1912, 64, 8 August 1912, 311; W. W. Macon in *Encyclopaedia Britannica* 31, 1922.
15. Fisher 1963, 144.
16. C. Hook to G. Verity, 5 March 1915.
17. *IA* 3 January 1924, 113.
18. *Mechanical Engineering* August 1927, 912; Borth, passim; Malborn c. 1956; *IA* 16 June 1927, 1792; *IA* 29 December 1927, 1828.
19. *IA* 16 June 1927, 1731–1740, Editorial, 1766.
20. *IA* 11 November 1926, 1357.

Chapter Seven

The Wide Continuous Hot Strip Mill

Strip, as its name suggests, is a long, narrow ribbon of thin steel. For generations this—or its predecessor, made from wrought iron—had been rolled for various lines of consumption. Some used it more or less in the form in which it was produced, as for instance in barrel hoops or cotton ties, or in fairly simple further-processed products, such as for welding into gas pipe. In the early 1890s, the maximum width and length of strip rolled was about 7 inches by 100 feet. Over the years both dimensions were increased, so that by 1925 strip in lengths of as much as 500 or even 1,000 feet was being produced in widths of up to 24 inches. Gradually it was realized that, if cut, wide strip could provide an alternative source for at least the narrower sizes of sheet. If it proved practicable to roll considerably greater widths, then it might be competitive over a large range of uses. An Ohio steel company had pioneered the continuous-sheet process at a Kentucky plant; a wide continuous-strip mill was another Ohio achievement, though it, too, was first installed in a nearby state. As always, a large number of people contributed to the development, but the breakthrough in strip was above all made by two men, Arthur J. Townsend and Harry M. Naugle.

Townsend was born in 1888 in the small west Pennsylvanian mill town of Apollo. By the First World War, he was employed by the Berger Manufacturing Company in Canton, Ohio. In 1916, along with two colleagues from Berger Manufacturing, Naugle and K. R. Jenson, he organized the National Pressed Steel Company, with a plant at Massillon, to make hot rolled strip up to 24 inches wide. In 1922, National was acquired by the Central Steel Company (later incorporated in Republic Steel) and Townsend, Naugle and Jenson moved on, this time acquiring the Columbia Steel Company of Elyria, a small town in the lake belt between Cleveland and Sandusky. Elyria had produced cold-rolled strip since 1902. In March 1926, Columbia Steel was reorganized so as to absorb the Forged Steel Wheel Company of Butler, Pennsylvania. This had been a subsidiary of the Standard Steel Car Company and produced over 10,000 tons a month of high-grade railroad wheels. Butler had no iron making capacity but contained a steel plant of ten large open-hearth furnaces. It was this steel capacity which persuaded the new owners of Columbia Steel that Butler should be the location for a new continuous mill they were planning. Involvement by the manufacturers of rolling mill equipment was vital to their success, an especially important part being played by F. C. Biggert, president of United Engineering and Foundry Company. Biggert cooperated wholeheartedly when Townsend and Naugle asked his company to design a strip mill to produce 30-inch-wide strip down to 12-gauge—too thick for use in car body shops without further reduction.[1] In turn, UE&F revealed its own researches in four-high mills and in the development of roller neck bearings. The new Columbia mill which the group of collaborators now built could roll strip up to 36 inches in width down to 14-gauge in four 4-high hot mills.

The Butler strip mill included a number of vitally important technical advances. At Ashland Tytus had found it necessary to part company with the old sheet mill principle that the configuration of rolls should be as near as possible perfect cylinders, and instead to work out a sequence of decreasing roll convexities in successive stands, so as to finish with a sheet of even thickness throughout. In contrast with his eleven 2-high and three 3-high mills and the batch production of relatively short lengths of sheet, rolling from a single slab, the four high stands which Townsend and Naugle installed turned out wide strip hundreds of feet in length. If wider material than 24-inch and lighter than 12-gauge was rolled in a two-high strip mill, it had been found that fundamental changes in mill design would be required, for otherwise neck friction would increase to unacceptable levels. The key to success was found in using two work rolls of small diameter, each backed by a much bigger roll, that is, a four-high mill. The larger outer rolls ensured that the smaller work rolls underwent very little deflection and were therefore able to roll wide strip to lighter gauges while maintaining uniform thickness. There were two other essential contributions to the success of the high-speed four-high wide-strip mill. One was use of large roller bearings on the necks of the rolls—bearings capable of withstanding great loads, of a style only available since the end of the war, and largely stemming from work done in central Europe. The other was a new level of sophistication in coordinating the speed of successive stands of rolling mills, for which the vital contribution was made by electrical machinery makers, particularly Westinghouse. This involved large, variable-speed electric motors, by which each set of rolls could be separately driven, and made it possible to avoid placing either too much or too little tension on the piece and to accommodate its rapidly increasing speed of movement as it passed through toward the run-out end of the mill. Too much tension would cause the piece to snap; too little would mean it buckled. All these technical advances came together to provide the basis for success—four-high mills and roller-neck bearings reduced friction and therefore the heat generated and together these gains reduced roll deformation and the amount of roll shaping required. In fact, the Butler mill solved a range of problems so conspicuously that, as *Iron Age* promptly recognized in a May 1927 major article on the new plant, "the sheet [*sic*] mill of tomorrow is in successful operation at the plant of the Columbia Steel Company, at Butler, Pennsylvania." In addition to its technical achievements, the Butler strip mill, like the Ashland continuous sheet mill, had an extraordinarily high productivity. The annual output of a hand mill with a crew of 10–14 men and working on lighter gauges of sheet was put in 1927 at a little over 6,000 tons; on heavier gauges, it might rise to 18,000 tons. At Butler mill the same size workforce could produce about 400,000 tons in a year.[2]

Having, as George Verity put it, "worked out" the new wide strip mill, the owners of Columbia Steel patented its essential processes just as the American Rolling Mill Company had done with Ashland. But as the latter's patents covered processes for rolling thin flat-rolled material more than 20 inches in width, it informed Columbia Steel that it suspected that its rights were being infringed. Apparently there were grounds for this belief, for, having studied the patents on the Tytus process, the leading partners of Columbia Steel decided to sell out. (They continued to pioneer, moving on to set up Rotary Electric Steel of Detroit and the engineering development firm of Naugle and Townsend Inc. of Canton. Arthur Townsend died in 1935 at 47.) In mid-July 1927, the American Rolling Mill Company bought the Columbia Steel Company. Now in charge of both Ashland and Butler, it had continuous mill capacity for roughly 850,000 tons of sheet or medium width strip a year. (Another source put its capacity at only 600,000 tons.)[3] Operations were soon discontinued at Columbia's Elyria works, but it was decided to build further on the successes of Butler. Before the end of the year, its new owners had acquired the patents of the United Foundry and Engineering Company covering four-high and continuous mills for sheet and strip, and in spring 1928 announced plans to spend $2 million at Butler for operations downstream from the hot mills, in annealing, pickling and cold rolling.[4] The urgency to undertake this expansion was in part a reaction to expanding market prospects, but also to initiatives taken by rivals, for by the end of 1927 three more wide-strip mills had been completed. By the end of the decade the total was seven. In tonnage terms, the statistics were dramatic. Continuous strip/sheet mills turned out 254,000 tons in 1926; three years later they made 1,758,000 tons.[5] With one exception, the new mills were confined to Ohio or its fringes.

Organized in 1912, by 1925 the Trumbull Steel Company of Warren, Ohio, possessed seven open-hearth furnaces, blooming and bar mills, and three hot strip mills rolling narrow material. It also controlled 13 sheet mills, two jobbing mills and 19 tinplate mills.[6] Like many other producers of thin flat-rolled products, Trumbull had expanded rapidly; in 1923 it had only 24 cold-strip stands, by the beginning of 1925 48, and by August of that year 61. It shipped 120,000 tons of finished steel in the third quarter of 1926, but then announced plans to increase its monthly production by fifty percent, at a cost of some $1.5 million and largely involving the widening of an existing 18-inch strip mill to 36 inches. In June 1927, when Armco and Columbia Steel were on the eve of a combination, Trumbull put its new strip mill into operation. It was now reported that it could roll 30-inch strip but that the mill would eventual be widened to roll to 36 inches. By then, it had allocated a further $1.25 million for improvements over the next few months.[7] On 25 November 1927, John A Topping, chairman of Republic Iron and Steel Company, and well acquainted with sheet products since his presidency of American Tin Plate a quarter century before, announced that his company had acquired Trumbull Steel. Republic brought to the amalgamation large coal and ore resources, iron making and a steel capacity beyond its existing finishing capacity; Trumbull had a share in a local blast furnace and some gas coal in Greene County, Pennsylvania, but above all it contributed its investment and hard-won know-how in continuous mill technology.[8]

The next strip mill had been decided on even earlier, but was not commissioned until some four months after the Trumbull mill. The company involved had already shown remarkable drive. At Weirton, on the Ohio in northern West Virginia cold rolling of strip began in 1913, and a few years the plant had been developed into a fully integrated operation. By 1925 Weirton Steel Company had a steel capacity of 570,000 tons; by the end of 1926 it reached one million tons. In spring 1926, the company "further broadened its construction program" to include a hot strip mill, which would now take precedence over a planned tube mill and extensions in sheet. The $5 million mill would double the company's existing capacity. By October 1927 the Weirton strip mill was at work, and although it began with material only 20 inches in width, it soon worked up toward a maximum of 48 inches. When rolling 38-inch strip, it could reduce it to 16-gauge—in both respects a considerable advance on Butler. As late as 1920, Weirton had made no raw steel; by 1929, it was expected to have a capacity of 1.2 million tons.[9]

In marked contrast to Armco, Trumbull and Weirton, the United States Steel Corporation had to defend rather than to forge a leading position in the sheet and closely associated tinplate trades. In the former in particular it had rapidly lost ground. In 1919 American Sheet and Tin Plate, its subsidiary for these products, accounted for 46 percent of national tinplate and 32 percent of sheet mill output. Seven years later its share was respectively 47 percent and 18 percent.[10] Capacity for these products was divided between a large number of small works and two bigger operations, the mills and open-hearth plant at Vandergrift, western Pennsylvania, and the AS&T mills at Gary. By 1926, the latter had 24 double black plate mills, with sheet and pair furnace equipment for 48 single hot mills. In September that year the USS Finance Committee approved expenditure of $8–9 million to install a hot strip mill to supply "breakdowns" of strip to the Gary hand mills, doing away with the need to reduce sheet bars in existing hot mills and thereby freeing them for more finishing work. It was reckoned this would increase the overall capacity of the mills by about 50 percent. The hand mill sheet went on to a tinning plant. In summer 1927, as the Trumbull and Weirton strip mills came into production, US Steel engineers were speeding up completion of the new Gary plant; before the year ended it was in operation as the nation's fifth wide hot strip/sheet mill.[11]

Commenting on the wider dimensions of strip now being rolled by strip mills, a December 1926 *Iron Age* editorial emphasized that this meant that "even up to 48 inches" wide strip was now used where formerly sheet had been required, as in the automobile industry Various types of mill were supplying the market—three or two high mills, tandem mills, continuous mills or even reversing mills. The author of the editorial believed that "out of this, with the passage of time and the spending of money, will come more or less standardization, with greater widths and lighter gauges."[12] The latter part of this assessment identified two of the deficiencies of the earlier strip mills, and thereby a continuing advantage of the older type sheet mills. Unfortunately for the latter, this proved a rapidly wasting asset. Within two years the commissioning of the new mill at Middletown brought the number of strip mills to six. The other five mills had been operating for a long time on schedules approaching capacity and at least three of them had been producing full-finished sheets of 22-gauge which involved investment in continuous cold rolling mills. H. L. Bodwell of Vandergrift, an AS&T operation which had no particular cause to champion strip mills, recognized "there are continuous mills in existence today capable of rolling 50 to 60 gross tons per hour into strips of 14 gauge or heavier, up to 48 inches wide and into 16-gauge, of 18 gauge up to 32 in. wide, at a saving in labor as compared to the similar roughing operations on old-style mills of $5 to $6 per ton." With 4-high continuous cold mills, this could already be successfully rolled down further into 22- or even 24-gauge. As the Chief Engineer of United Engineering and Foundry put it at this time, the four companies operating these six mills "are large, well-managed, and are accustomed to making profits."[13] In May 1927, *Iron*

Age, after reviewing the Butler mill, recognized the scale of the challenge the new technology posed for the whole sheet industry: "It does not take much calculation to realize that only a limited number of single continuous roll trains, each capable of making 150,000 to 300,000 tons per annum, could satisfy the country's needs of 4,500,000 to 4,700,000 tons. Meanwhile, owners of old mills will need to do no small amount of planning if they desire still to be numbered among the purveyors of rolled steel."[14] The threat was a real one, and its outcome would eventually prove lethal to the older technology. Even so, for many years there would be mitigating circumstances, and ways would be found of alleviating the impact.

Even before Butler, it was recognized that the strip/sheet industry had been thrown into turmoil. Yet it was still uncertain which way it would go. This was very much a time of transition, with diverse opinions as to the gauge and width continuous mills could attain, about their profitability and therefore about the role which was left for hand mills. The technology and the economics of the sheet mill and the strip mill industries were completely different. They used steel in different forms—bars in the case of the first, "bar plate" at Ashland and slabs in the strip mills. As most sheet mills bought their bars, they had to be careful in monitoring the chemical analyses and physical properties of what they purchased. Strip mills usually had their own steel supplies, but the process imposed exacting requirements in terms of low-carbon, low-phosphorus, deep-drawing steels, and, as Charles White of Republic Steel later summed up, in addition to outlay for the strip mill there might be various extra costs. Making slabs for the strip mill required large ingots and therefore sometimes made obsolete existing soaking pits and in many cases the blooming mill as well. The open-hearth steel had to be of good quality, and this might need better pig iron. The contrasting economics of the two methods were summarized in a paper given by Leon Cammens, a consulting engineer, to the November 1928 Chicago meeting of the Iron and Steel Division of the American Society of Consulting Engineers. Cammens reckoned that a sheet mill could be built for $250,000, whereas a wide strip mill would cost at least $2 million and would therefore have interest charges of the order of $120,000 a year as well as high depreciation costs. But taking into account all the ancillary plant, the capital outlay for the two technologies for a given strip/sheet output was more or less comparable.[15] It was true that in special cases the economic assessment might be different from this. A case in point was the Gary installation, commissioned in November 1927. American Sheet and Tin Plate had 48 hand mills there which coped with both roughing and finishing operations. Increasing capacity as desired would involve either installing 48 more hand mills or putting in a hot strip mill to undertake the roughing work, freeing that half of the hand mill capacity which had till then done this work to become finishing mill capacity. It was shown that 48 new hand mills would cost, with necessary ancillaries, $18 million, but provision of a strip mill and conversion of present roughing capacity to finishing could be done for $12.6 million. Additionally, it was reckoned that with a strip mill employment could be reduced by 200, thereby cutting labor costs per ton of hot-rolled product by about $2.5.[16] More generally, as a labor-intensive process, sheet mills had costs for labor which Cammens put at $5.50–$6.00/ton of sheets, whereas the highly mechanized strip mill could obtain extremely low labor charges per ton of output—if operated at full capacity.[17] But whereas sheet mills were flexible as to widths, gauges and variations in demand, strip mills could only provide good returns if running more or less full on large tonnages of relatively few specifications, and therefore not needing frequent changes of rolls. If operations fell to a low level, their overheads might prove crippling. As *Mechanical Engineering* had recognized in the immediate aftermath of the first news of the Butler mill, "A continuous mill is substantially a tonnage producer, and will pay only if used as such."[18] Much was made of the "flexibility" of old-type mills. Though the automobile industry was a large volume buyer, "the very fact that there remain large numbers of small consumers demanding wide varieties of gauge, size, and finish will save those outside the new movement from becoming panic stricken."[19] For a time it also seemed that hand mills might continue to be needed because of the early inability of strip mills to roll wider and lighter strip-sheet. All of these factors, favoring the survival of the old-style mills, real though they were, proved to be wasting assets.

But at the same time, in response both to market opportunities and to the new challenge, the old-style sheet mills significantly improved their own operational efficiency. They separated roughing and finishing operations, conducting them in separate stands. Working conditions were improved and labor costs reduced by introducing continuous pair furnaces, and mechanical doublers and means of handling both sheet bars and packs. Eventually they would introduce three-high roughing mills and fully mechanized two-high finishing mills. At an early stage of these improvements, R. J. Wean of the Aetna Standard Engineering Company of Youngstown, and therefore an interested party, painted a rosy picture of the prospects: "When all of the economies of this nature have been accomplished in what we shall call the 'conventional-type' sheet mill, it will be extremely difficult for wide-strip-sheet producers to manufacture at a lower cost than the ordinary sheet-mill producers." In fact, nothing was to prove able to equip the old-style industry to match the quality or the economics of the strip mills. Wean's optimistic prognostications were sadly wide of the mark.

Though strip was supplanting sheet as a material for automobile fenders by the early 1920s, for a time the hot strip mills were not able to supply much of the material used for car bodies. The first "wide" hot strip mills could not roll widths comparable with those of sheets rolled on hand mills or the thinner (or "lighter") gauges which were in rapidly increasing demand. In spring 1927 *Mechanical Engineering* reckoned strip mills could roll only up to 42-inch widths. However, as wider strip became available, the main consumers, and especially the motor industry, began to demand ever wider sheets. The customer as always was setting the pace and asking for more, while not being willing to pay more for it. (In 1928, Cammens referred to talk in the sheet trade of educating purchasers to pay more for these products and so eliminating "ruthless" buying. As he neatly summed up, "This is the old story of the mice that met and decided that all cats should submit to having their teeth drawn.")[20] For wider material and lighter gauges—or to meet smaller orders, for which the massed capacity of a strip mill was less suitable—it was still necessary to resort to the old-style sheet mills. Up to 1925, the widest sheets ordinarily rolled by sheet mills were 48 inches; by 1930, they could produce sheets 72 inches wide.[21] However, during the late 1920s, the width limitations of the strip mills were also pared away. Wider mills were built, such as the 56-inch mill at Middletown and at Steubenville a 66-inch mill owned by Wheeling Steel. Existing strip mills were reconstructed to roll wider strip, Butler, for instance, increasing from 36 inches to 48. The strip mills also made headway in regards to thickness. Early continuous mills were confined to thicker gauges of sheet or strip, at Ashland and Butler no thinner than 16-gauge. The Gary mill rolled breakdowns for its finishing mills down to 18-gauge, but the limit at Middletown was initially 14-gauge.[22] The limitations on gauges were effectively removed by the introduction of cold reduction mills. The first four-high cold mills for sheet products were in operation in 1926, when total capacity was about 150,000 tons. In spring 1928, Armco announced it would spend $2 million at Butler on annealing, pickling and cold rolling plant.[23] The cold rolling of hot-rolled sheet and strip grew rapidly. It brought important further benefits, for the quality and finish of the product was superior to that which could be achieved in hand mills. In short, together the hot strip mill and the cold reduction mill quickly swept away the technological defenses of old-style sheet mills, so that it became clear that as the strip mill advanced a gradual erosion of the outlets for hand mills was inevitable. If the more or less steady increase in demand, which through the last years of the 1920s had so far supported both the new and the old technologies—and in 1929 even allowed the old-type mills to turn out a record tonnage—slowed or was reversed, the hand mills would face a severe crisis.

Another important, though much less momentous, issue concerned rights to the new technology. Butler had been sold to the American Rolling Mill Company partly because its owners did not want the uncertainty of contesting its patents. Weirton Steel contracted independently with United Engineering and Foundry for its new strip mill, and US Steel engineers designed the strip mill installed at Gary. In late summer 1927 Weirton took out a license to use all the American Rolling Mill patents for strip/sheet and strip.[24] Almost two years later, the US Steel finance committee approved an arrangement whereby the Corporation would pay American Rolling Mill $50,000 a year for patents already in use in Gary works and a royalty of 30 cents a ton of products rolled by this equipment at any other of its mills.[25] By October 1929, licenses to use American Rolling Mill patents had also been taken out by Wheeling Steel, Youngstown Sheet and Tube, Republic and Gulf Sates Steel. By December 1929, as things turned out an unpropitious time, Otis Steel also took out a license and was reported as planning to spend $10 million on a Cleveland strip mill.[26]

By fall 1929, seven hot strip mills were in operation. Their rated capacity was over 2.6 million tons. The capacity of four-high cold reduction mills for either sheet or tinplate was about 800,000 tons. Over the next six years, the number of hot strip mills would increase to 20 and their capacity to 8.4 million tons. By the end of that period, 4-high cold mill capacity reached 4.4 million tons.[27] Between the two dates stretched the dreary waste of the depression and years of hard struggle involved in emerging from it.

NOTES

1. J. Malborn in correspondence; see discussion of L. Cammens paper to Iron and Steel Division of American Society of Engineers, Chicago meeting on strip mill in autumn 1928 reported in *ITR* 21 December 1928, by J. W. Sheperdson; R. J. Wean of Aetna-Standard Engineering Co, L. Jones of EW Bliss Engineering of Salem and H. L. Bodwell, Assistant District Manager of AS&T, Vandergrift.
2. *IA* 19 May 1927, 1435–1439; see also for a review of the whole field F. C. Biggert, "Developments in 4 high Rolling Mills" in *TAIME* Vol 49–50, Part 2 1927–1928; Iron and Steel section.
3. *BFSP* August 1927.
4. *IA* 15 December 1927, 1681, 26 April 1928, 1152.
5. *AISE* 1941.
6. *IA* 3 September 1925, 6470.
7. *IA* 27 August 1925, 562, 21 October 1926, 1175, 23 June 1927, 1852, 15 September 1927, 693.
8. *IA* 1 December 1927, 1555.
9. *IA* 8 April 1926, 1007, 17 May 1928, 1427.
10. Ford, Bacon and Davis *Report 7* on US Steel.
11. *IA* 16 September 1926; *ICTR* (UK), 3 June 1927, 890.
12. "The Strip-Sheet Evolution" in *IA* 9 December 1926, 1.

13. H. L. Bodwell and L. Johnson in discussion of Cammens, 12, 13.

14. *IA* 19 May 1927, 1462.

15. See discusson of Badlam paper in *AISI Yearbook* 1927, 418,419; C. M. White, "Technological Advances in Steel Products," *AISI Yearbook* 1937, 110, 113; Cammens 1928.

16. Hogan, 856.

17. Cammens, 3,6

18. Quoted *BFSP* June 1927, 266.

19. C. W. Bennett of AS&T in discussion of Badlam, quoted *IA* 3 November 1927, 1234; *IA,* editorial 19 May 1927, 1234.

20. Cammens, 9; see also "Sheet steel for Automobile Bodies" in *BFSP* February–November 1928.

21. Lawrence 1930, 214.

22. Hogan passim.

23. *IA* 26 April 1928.

24. *IA* 1 September 1927, 585, 5 January 1928, 107.

25. USS Finance Committee mins 9 April 1929.

26. *IA* 3 October 1929, 921, 5 December 1929, 1560, 12 December 1929, 1625.

27. *IA* 2 January 1936, 89.

Chapter Eight

Investment in Adverse Times

The 1930s

Even before the end of the prosperous 1920s, some well-informed people were questioning the wisdom of the huge investments represented by the new continuous sheet and strip mills. For instance, in autumn 1928 John Sheperdson, Chief Engineer of Morgan Construction Company, commented that the sheet-mill industry had "plunged into a new process almost regardless of cost, allured by the potential advantages in sight."[1]

The important 1927 paper in *Mechanical Engineering,* quoted earlier, had concluded that "it would appear that some seven continuous mills could take care of all the sheet rolled today in the United States."[2] Ten years later, the director of the Pittsburgh Bureau of Business Research summed up what had happened since then in somber terms: "The past ten years or so may be looked on as years of transition for the American iron and steel industry—transition from youth with its vigorous growth and rapidly expanding markets to the stage of economic maturity, attended by slow growth in productive facilities and intense rivalry for slowly expanding or stationary markets. Profits no longer accrue more or less automatically, but must be garnered as the fruits of competitive efficiency."[3] Producers of sheet and strip must have grimaced or smiled wryly when they read about "slow growth in productive facilities," for despite deep and prolonged depression, this had been a decade of feverish construction of hot strip mills.

Until 1927, Ashland was the only continuous mill producing wide flat-rolled products; by the end of 1929 there were seven hot strip mills and installed capacity of continuous 4 high strip/sheet mills was a little more than 2.9 million net tons. Over the next seven years, 11 more strip mills were added, and capacity reached almost 8.9 million. By the end of 1937, there were 21 continuous mills which could roll sheet or strip of 38-inch width or greater. Another 8 mills were built by 1940, when total capacity reached

16.0 million tons.[4] (This figure includes one "merchant" strip mill, that built by US Steel at Youngstown, but other two mills—Homestead and South Chicago—are excluded as being effectively light plate mills rather than strip mills.) Given the state of the national economy, it was difficult to avoid the dangers of overexpansion, but no major company could risk missing the opportunity to be involved in the section of the steel industry which was seen to hold the best prospect of long-term growth. The apparent rapid rise in the average capacity of mills was deceptive, for in both of the later periods some of the advance was the result of the improvement of existing mills. As far as output was concerned, strip mills were held back in the early thirties by the exceptional decline in trade, and later by lack of adequate supporting plant and equipment, for expenditure on this was skimped during the bad years.

One interesting feature was a change in the pattern of ownership of strip mills. Except for the USS Gary mill and that installed at Steubenville by Wheeling, all the units completed before the Great Depression were owned by relative newcomers to the list of important companies. But from that time onward long-established major steel firms were more prominent. This reflected the changes of the times in the steel market. In 1929 sheets, plate and strip made up 33.6 percent of all hot rolled finished products; by 1936 their share was 38.0 percent. For each of these products output in 1936 was higher than in 1929; for only three other of 19 categories of steel product listed by the AISI was this so, and in tonnage terms all of these other categories were much less significant. For strip mills—including the two continuous light plate mills—the changes were dramatic. The 1936 output of all hot rolled products was almost 18 percent less than that of 1929; production by hot strip mills was 375 percent larger than seven years before. Despite the Depression, only in four years did strip mill output fall back from levels previously reached—in 1930, 1931, 1932, and 1938. The strip mills turned out only 3.83 percent of all hot rolled products in 1929 but 34.3 percent in 1940. The Great Depression brought out vividly the lack of balance and even more the unsuitability of the product lines of many of many long-established producers to meet the circumstances of the new age. As *Fortune* pointed out in December 1935, the two leading moneymakers through the depression years, National Steel and Inland Steel, had by then 68 percent and 51 percent respectively of their capacities in bars, sheet and strip, the main outlets for steel for the automobile industry. In sharp contrast, for US Steel and Bethlehem Steel the proportion in these products was 31 percent and 28 percent, respectively.[5] At the beginning of that year, four major companies had only commissioned one strip mill between them. US Steel had operated the Gary mill from 1927 but Bethlehem, J&L and Youngstown Sheet and Tube still had no strip mill in operation. After being left behind in the race to move from a traditional range of mainly heavy products to new, consumer-oriented lines, they now had to make every effort to catch up. Over the next five years, these four companies accounted for two thirds of the new strip mill capacity.

Another leading theme of the thirties, also linked with strengths and weaknesses in corporate policies, was the movement of the center of gravity of the new industry from the landlocked locations in which it had begun in the 1920s to situations better for low-cost steelmaking along the shores of the Great Lakes. Of the over 4.2 million tons of capacity existing in 1940 at strip mills first set up before the end of 1929, 88 percent was in the interior; of the 7 million tons at mills installed from the beginning of 1930 to the end of 1936, only 23 percent. Almost 31 percent of the 4.8 million tons installed in the three years to the end of 1939 was at interior plants. It must, however, be noted that well over one quarter of the capacity in those interior plants built in the 1920s had been added after that period as they were improved or rebuilt.

One of the most favored districts for strip mill construction was Chicago. US Steel's American Sheet and Tin Plate built the first strip mill in the area in 1927, and this was followed by other USS mills in 1935 and 1936. By 1940 Gary shared with the National Steel works near Detroit (Great Lakes Steel) the distinction of having, at 1.5 million net tons, the largest strip mill capacity of any works.

Not far short of these leaders was a company for which the switch to thin flat rolled steels marked an impressive and well-timed change in direction. Inland Steel of Indiana Harbor had rolled sheets on two high mills from 1908, but as late as 1930, 80 percent of its capacity was in heavy products. It now built a strip mill wider than any yet installed and a large cold reduction mill. Commercially, Inland was fortunate in that these major extensions were made at a time of deep depression, when construction costs were low. As no strip mill of comparable or greater width was built for at least three years, the company was able to build up a valuable connection with customers, which it carefully fostered. The Inland mill made a profit in every year except 1934, and by the end of the decade it was reckoned to have been the best moneymaker of any of the wider strip mills.[6] In 1932, Indiana Harbor first produced tinplate. Five years later, Inland brought into production a second strip mill, this time 44 inches wide, on a site reclaimed from Lake Michigan. Only 40 percent of the company capacity was now in heavy products. Its commercial success was helped by the fact that three quarters of its deliveries were made within a distance of 150 miles from the plant. *Fortune* in October 1938 provided a detailed breakdown of the Inland costs of production all the way from mining and iron making to the rolling of the finished cold rolled strip.[7]

Table 8.1. Capacity of hot strip mills, 1940, by district and date of construction (thousand tons)

	*Interior**	*Lake Erie*	*Chicago*	*Other*
To end 1929	3,758	—	492	—
1930 to end 1936	1,618	3,157	1,714	470+
1937 to end 1939	1,478	900	1,232	1,176+

* Interior is between River Ohio and Great Lakes

Although it had made an outstanding positive response to a crisis, the challenge for Inland was simplified by the fact that it possessed only one integrated works, and this was in a prime location for assembling materials and marketing. In other cases, long-established heavy steelmakers found it more difficult to decide on the location for a strip mill. Building in a time of recovering trade, they also had to pay more for construction. In 1929, the National Steel Corporation bought 1,000 acres of land straddling the Burns stream near Valparaiso, Indiana. Hopes of a major integrated works there to parallel its Ecorse works were frustrated by the Depression. In 1939, when the state of Indiana appropriated $50,000 to develop a harbor on Lake Michigan, National was expected to begin to build either that year or in 1940. In fact further development was delayed for another 20 years.[8]

In 1930, Bethlehem Steel made a bold attempt to widen its range of choices for future production by means of a merger with Youngstown Sheet and Tube, operator of important works in both the Youngstown and the Chicago areas. The bid was defeated in the courts, and Bethlehem eventually chose what seemed the most rational location for its first strip mill by installing it at Lackawanna. Two years later, it built its second mill at Sparrows Point. Despite these two major projects, by 1938 only 23 percent of Bethlehem capacity was in sheet, strip and tinplate as compared with 44 percent at Inland Steel.[9] The pattern of development at other major companies threw some doubt on the long-established conventional wisdom that in marked contrast to hidebound steelmakers in the Old World, those who controlled the American industry let nothing interfere with the pursuit of commercially optimum solutions. A number of old-established companies showed a partiality for investing in the new technology in their old, headquarters locations rather than where there would appear to have been greater advantages.

The biggest markets for strip mill products were in the consumer durable industries. Though widely spread, these were above all concentrated in the northwestern parts of the manufacturing belt. The main centers of steel processing there generated large tonnages of scrap, and this was usually available at prices well below those in the leading interior steel centers. Locations along the shores of the Great Lakes were well placed not only to enjoy market and scrap advantages but also to lay at break of bulk points for the iron ore and coal which fed the blast furnaces. Most of them offered favorable site conditions for the space-extensive hot strip mill, in marked contrast with the often cramped physical settings of the interior metallurgical centers, in which long years of piecemeal expansion had also further compromised the possibilities of a good plant layout. One company responded very positively to these opportunities. When formed in spring 1930 as an amalgamation of a number of long-established companies, the new Republic Steel Corporation already controlled one strip mill, at Warren, Ohio, which it acquired with Trumbull Steel in 1928. In 1933, 81 percent of Republic's raw steel capacity was within 60 miles of Youngstown, which contained its head offices. In April 1934, a merger was agreed with Corrigan-McKinney of Cleveland, and although arrangements were delayed by court hearings, by September 1935 it had been approved by shareholders. Republic moved its head offices to Cleveland, where by 1937 it had built a 98-inch hot strip mill, the widest yet, with more than twice the capacity of its Warren mill.[10]

Youngstown Sheet and Tube was less decisive. Since 1923 it had owned the Indiana Harbor works originally built by The Steel and Tube Company of America. In its eventually frustrated bid to merge with YS&T in 1930, Bethlehem Steel made clear that one of its main aims was to acquire the Chicago plant in which it would concentrate further investment. When Youngstown Sheet and Tube responded to the necessity to build its own strip mill, it opted for a less imaginative strategy. In 1935 it opened a 670,000-ton strip mill at

Table 8.2. Finished steel capacity, 1933, by company and main categories (percent of total)

	*Heavy products%**	*Sheet and strip%*	*Tinplate%*	*Other%+*
New or reconstructed companies				
National Steel	c.10	61.8	14.0	14.3
Inland Steel	39	24.3	—	24.8
Republic Steel	c.3	24.7	c.3	c.70
Long-established "heavy" companies				
Youngstown Sheet &Tube	13.5	9.5	c.4	69.2
Jones and Laughlin	26.5	—	<6	67.4
Bethlehem Steel	55.5	<5	<5	<36
United States Steel	37.3	c.10	c.5	46.4
Average US output 1923–33	32.5	20.3	5.6	41.6

* Rails, structural shapes, and plate
+ Bars, pipes and tubes, wire rods
Figures do not total to 100%, especially in the case of Inland Steel
Based on *Fortune,* September 1933

its Campbell works, Youngstown. A year earlier it had made a tentative move at its Chicago plant. A 72-inch mill was opened there, but this was a reversing, not continuous, mill, and its capacity was only 240,000 tons. Not until 1939 did YS&T commission a large strip mill at Indiana Harbor, with a rated capacity of 672,000 tons.[11]

In the mid-1930s, major steel companies were either installing their first strip mills or making major extensions in plants already built. In 1936 at Gary, US Steel commissioned the first 80-inch hot strip mill, the Lackawanna mill was almost as wide, National Steel built its second strip mill of the same size, and Jones and Laughlin were at work on the widest mill to date in Pittsburgh. The next year, US Steel, Bethlehem, Republic, and Youngstown Sheet and Tube planned to spend a total of $94 million for hot and cold strip mills.[12] Meeting such challenges, other companies had to put in new facilities to produce the wider dimensions now being demanded. For any firm which had not begun strip mill construction, it was high time to bestir. An interesting case of the commercial necessity felt by companies to hold their place in the trade was that of Granite City Steel.

From the 1890s, this company had produced plate and sheet. By the early 1920s, it had an integrated works with a capacity of 400,000 net tons of steel

Table 8.3. Regional distribution of sheet and tinplate capacity, 1922, and strip mill capacity, 1940 (thg ts)

District	*End 1922 sheet*	*End 1922 tinplate*	*1940 strip mill*
Eastern	—	—	—
Philadelphia	143	93	672
Johnstown	—	—	—
Buffalo	80	—	806
Cleveland	211	27	1,320
Pittsburgh	930	664	2,290
Valleys	754	471	1,744
Wheeling	624	601	1,660
Columbus	867	70	1,160
Detroit	—	—	1,931
St. Louis	132	90	470
South	—	—	504
Chicago/Ill/Ind	538	266	3,438
West	—	—	—
USA Total	4,277	2,282	15,995
No. of works	78	34	26
Plant average cap	54.8	67.7	615.2
No. works with steel cap.	30	8	26*

* Steel for the McDonald strip mill was carried from the USS Youngstown works
Based on *IA* 1 January 1925, pp. 10–17, and J. Ess, *The Strip Mill* (1941)

and 220,000 of sheet, light plate and black plate. It had galvanizing and tinplate plant. Decision to build a strip mill was taken in 1935, but the company had to scrape to find the necessary resources. One means of cutting capital outlay was to use an existing 100-inch three-high plate mill to provide the roughing stands for the strip mill—though this reduced output. The Granite City 90-inch strip mill, commissioned in 1937, and the Otis mill in Cleveland had the smallest capacity of the wide-dimension mills of the 1930s.

In this period, the industry began to spread, though slowly and on a modest scale, from the Northeast. Before the Wall Street crash, Gulf States Steel of Gadsden, a small integrated producer which concentrated on bars, hoops, bands and so on, and which in the 1920s had no involvement in sheet products, took out a license to use the American Rolling Mill patents. Presumably in large part because of the Depression, it did not build a strip mill. In 1937, Gulf States was acquired by Republic Steel. As early as March 1930, the finance committee of US Steel recommended expenditure of $8.1 million at its Fairfield, Alabama, works for a new open-hearth furnace and blooming mill, additional sheet mills and a strip mill.[13] This too was delayed by the economic circumstances—whereas further developments at Gary and the McDonald works (Youngstown) went ahead. By 1937, a 48-inch strip mill was at last completed at Fairfield. Initial annual capacity was 335,000 tons, but by 1941 this was expanded by about 50 percent.

NOTES

1. R. J. Watkins in preface of CR Daugherty, M. G. de Chazeau and S. Stratton, *The Economics of the Iron and Steel Industry* 1937, vii.
2. Ford, Bacon and Davis. *Report 200 part 2* 1938, *Report 9* 1936; *AISE* 1941, 5.
3. *Fortune,* December 1935, 81.
4. G. A. V. Russell 1938.
5. *BFSP* March 1932, 268; *Fortune* October 1938, 44, 46 (which includes detailed costs of production), July 1938, 98.
6. *IA* 19 January 1939, 633, 16 March 1938, 84, 21 December 1939, 57.
7. Russell 1938, 30, 31.
8. *Fortune,* September 1933, 120.
9. *IA* 2 January 1936, 85; *AISE* 1941, 5.
10. *IA* 7 January 1937, 70, 73.
11. *IA* 1 January 1925.
12. *BFSP* December 1949, 84, 85; *I&S Engineer* February 1969, 77.
13. USS Finance Comm. mins 25 March 1930.

Chapter Nine

Contrasts in Development Planning

Policies in the Pittsburgh District and Opportunities and Initiatives in Michigan

In the mid-1930s an interesting discussion was held at a meeting of the American Iron and Steel Institute. D. M. Petty of Bethlehem had presented a paper on "The Economic Importance of the Replacement of Obsolete Equipment in Steel Mills." In response, G. C. Kimball of the Illinois Steel division of US Steel contrasted the hit-and-miss procedures of the past with more rational modern approaches: "In the old days, extensions and improvements were largely made wherever the producing company had a large plant, regardless of the geographical location of the market. Today, before any improvement of consequence is made, it is the practice to make a market study to develop the probable demand, and the most economical point of location for such a facility, not only from the standpoint of manufacturing the product at a low cost, but particularly with respect to the final delivered cost."[1] Some of the strip mills proved the inappropriateness of such a reasoned procedure.

By the end of 1935, 16 strip mills had been installed with a combined annual capacity of 7 million net tons. For many decades, the Pittsburgh district area had contained the largest share of national steel capacity. In 1922 it had also contained a little under a quarter of the nation's sheet, light plate and black plate capacity. Thirteen years later, not one of the large new strip mills had been located in this district. (To this general statement there was one, partial exception, that of the small, specialized Allegheny Steel company at Brackenridge, on the Allegheny River.) This remarkable anomaly eventually had to be remedied, but after being slow to act, the two largest Pittsburgh companies, Jones & Laughlin and the Carnegie-Illinois subsidiary of US Steel, proved reluctant to break away from their old locations. In both companies, rational long-term planning seemed to give way to accommodation to existing circumstances and expediency.

In the 1930s J&L possessed two main works, the long-established complex of iron- and steelmaking plant and rolling mills which straddled the last miles of the Monongahela as it flowed into Pittsburgh, and the thirty-year-old operations on a new site down the Ohio at Aliquippa. By 1930, Aliquippa was a 1.4-million-ton plant; raw steel capacity in J&L Pittsburgh works was 2.0 million tons. The company suffered heavy losses in the Depression and failed to keep its ablest men, notably Tom Girdler, who in 1929 left to join Republic. However, before the end of 1935, J&L announced a $40 million program for new plant, improvements and expansion, including a $25 million hot strip mill with an annual capacity of 600,000 tons. Aliquippa was judged sufficiently diversified already, but the Pittsburgh works was not, and it was the location chosen to receive the new mill. By 1937, after a great deal of demolition of property and realignment of transport facilities, a 96-inch strip mill had been installed. (The J&L site had been hemmed in by the Monongahela on the south and by Second Avenue on its north side. An office block, pig casting machines, billet yard and three large cranes had to be removed.)[2] Steel was supplied from across the river. Capital outlay on the new mill was low as compared with the venture on which US Steel was soon to embark, but the latter seems to have included more ancillaries. Five years later, when it absorbed Otis Steel of Cleveland, J&L inherited a hot strip mill commissioned in 1932. It was comparatively small but much better located than the Pittsburgh works.

The situation at US Steel was more interesting and, fortunately, was fully documented. After its own engineers had spent seven years investigating what was needed to overhaul all its operations, their findings were recognized as so far-reaching that it was decided to bring in outside expertise. In autumn 1935, the consultants Ford, Bacon and Davis began a study of the whole of the Corporation's operations, and this eventually resulted in some 200 volumes of analysis and recommendations. In August 1937, in Report 100, the consultants dealt harshly with the Corporation's "Organization." Focusing on the question "Has the Corporation kept abreast of its market demand?" Ford, Bacon and Davis concluded, "A general review of the three and a half decades constituting the Corporation's record fails to confirm that it has met or anticipated as it might have, in our judgement, the reasonable demands of certain of its most important customer requirements. . . . We refer particularly to 1. Its failure to introduce years before it did the continuous mill, though this mill was presented first for its consideration."[3]

American Sheet and Tin Plate had an almost commanding lead in the output of thin, flat-rolled products in its early years. For long it was one of the Corporation's most profitable subsidiaries, but it had rapidly lost its share of national sheet capacity—in 1915 35 percent; by 1936 only 13 percent. In the boom year of 1920, net income was $21.3 million on gross sales of $212.2 million; by 1935 as the industry began a strong recovery from depression gross sales were above half the 1920 level at $113.9 million but net income was only $3.55 million. That year only 141,000 tons of AS&T output of 510,000 tons of regular sheet was delivered to the automobile industry. The share of US Steel in this sector of the industry had been relatively small and had been declining, whereas industry-wide capacity had been expanded well beyond what the consultants estimated to be the "normal" market—in short, any new strip mill would have to force its way into a situation of overcapacity.[4] Within the flat-rolled products sector, USS was more prominent in long-established than in hot strip products. It had taken an early part in the strip mill industry at Gary, but its involvement with the new technology was relatively small. The leading executives were fully aware that it was losing market share.

The question as where to build any new USS strip mill capacity also became a lively issue. Just before J&L announced its new Pittsburgh mill, American Sheet and Tin Plate revealed it too was thinking of building a strip mill in the area. This was not commissioned until 1938, a year after the J&L mill. The delay was due to an extended controversy within US Steel about the best place from which to serve the demand of the East and Northeast, areas beyond easy reach from its large capacity at Gary. In this debate a corporate bias in restricting choice of location was fully exposed.

In fall 1928, the decision was made to spend some $3 million to reconstruct Vandergrift to take advantage of improvements in sheet rolling, but the trend there was to a more specialized product.[5] Most of the Corporation's steelworks in the Pittsburgh area seemed to have fallen away from their earlier leadership in efficiency. Now the management of Carnegie-Illinois regarded it as essential that the situation of these works should be maintained, whereas experts from outside believed this was not a logical starting point

By 1937, Ford, Bacon and Davis were generally highly critical of the local works. For instance, they decided the number one open-hearth plant

Table 9.1. US Steel Corporation output of finished steel, 1936, as proportion of industry (percent)

Light flat-rolled		*Heavy and other*	
Black sheet	18.0	Rails	54.8
Tinplate	32.5	Heavy structural	46.4
Black plate	35.6	Plate	36.8
Hot-rolled strip	20.7	Wire rod	39.4
		Pipes and tubes	28.3

Source: FB &D Rep. 200, April 1938

at Homestead was unable to produce economically under conditions at the time, and indicated that the number two open-hearth plant "housekeeping . . . is exceptionally poor." Except for its ten-year-old structural mill and a new plate mill, equipment in its rolling mills it "is all old and requires replacement by modern capacity." At Duquesne, with few exceptions, the facilities were "old and in poor condition," so much so that the consultants recommended "a study should be made of the feasibility of placing the capacity of this plant in some other locality, with equipment capable of meeting market conditions profitably." For several years during the Depression the Clairton works had not operated, and general plans had been made to abandon it. "The policy seems to have been to defer all maintenance, and the equipment, buildings, wiring and piping very markedly reflect that determination." Edgar Thomson works was a striking exception. Here too there was much old plant, but "the works has been intelligently supervised in every respect, and . . . the maintenance in particular, has been carried out in such a manner that it might well serve as a model for the district."[6] ET's great misfortune was that its main product had been rails and sheet bars, for both of which, though for different reasons, demand had fallen sharply and could not be expected to return to the old high levels. In such circumstances it was obviously very attractive for Carnegie-Illinois planners to see ET as a possible source for the steel slabs which might be rolled down on a strip mill built at one of the other long-established and possibly redundant US Steel plants in the area.

In August 1934, US Steel's own Flat Rolled Products Committee had outlined a program of "rehabilitation" and "improvement" for sheet and tinplate. Step one involved expansion in the Chicago area. This was carried out with the commissioning of a 326,000-net-ton strip mill at Gary in 1935. The existing 28-inch two-high mill there was reconstructed as a 36-inch four-high mill. In the same year, a 42-inch strip mill was installed at McDonald works near Youngstown. The third major step believed necessary by the 1934 committee involved an overhaul of sheet and tinplate production in the Pittsburgh area. Eighteen months later, on 7 February 1936, it recommended expenditure of $60 million there. Ford, Bacon and Davis took the lead in challenging whether it was wise to build a mill in Pittsburgh.

By spring 1936, the local management of Carnegie-Illinois was convinced that work should begin at once on a strip mill at Clairton, but on 1 April, Ford, Bacon and Davis, in what they recognized was only an interim report, made clear their reservations about this plan. In their opinion no expenditure should be undertaken until a market analysis of the eastern area was completed.[7] In June, they recognized that the eastern part of the manufacturing belt made up a major market for sheet and strip products and that US Steel commanded only a small part of it. In 1935, for instance, the fourteen states of this area (from New England to the western edge of Ohio and to the southern borders of Maryland, West Virginia and Ohio) consumed 1,818,000 tons of uncoated sheet, of which US Steel supplied only 143,000 tons or less than 8 percent. To improve its position, the Corporation needed to increase its capacity, but Pittsburgh was not necessarily the right place at which to do so. Indeed: "Based on the lowest assembly, production and delivery costs, our studies thus far raise serious questions as to Clairton being the best location for such proposed new sheet mill facilities." They pointed out that the USS plant at Lorain was better placed to serve eastern markets and had the advantages of a lakeshore location: "Therefore we cannot approve of the proposed $60 million expenditure at Clairton until the Lorain situation has been explored further to determine what excess capital cost, if any, the development of Lorain would entail, and more carefully consider other factors." They noted plans for a new works had been produced by the company's steel engineers and were not accompanied by reports from sales. As no adequate market data had been provided, they asked for sufficient time to procure it.[8]

By September 1936 a report from Robert Gregg of US Steel confirmed that its engineers recommended Clairton. At more or less the same time a joint study of the relative merits of Clairton and Lorain was made by B. H. Lawrence of US Steel and H. E. Wachter from FB&D. They indicated that a Lorain development would require coke ovens, new blast furnace and open-hearth capacity, Bessemer and duplexing plant, and an additional slab mill. The estimated capital outlay at Lorain would be $72.29 million. If at Clairton the existing blast furnaces, open-hearth plant and most of the rolling mills, except for a 22-inch structural mill and 18-inch bar mill, would be closed, the site would be cleared, and an 80-inch strip mill and cold reduction mill would be installed. Slabs to supply the mill would be transported from Edgar Thomson, where the open-hearth plant would have to be enlarged and a slabbing mill built. Capital investment required at these two works was estimated at $59.62 million. In terms of production costs, Lorain would be the cheaper by $828,407 a year if calculated on a "tariff" freight rate basis or $634,151 on an "integrated" rate basis. In costs of distributing finished products, Lorain would be cheaper by $269,000. Lawrence and Wachter concluded, "The economic success of the undertaking is, therefore, in doubt and under these circumstances, the program should be deferred until it can be demonstrated that a reasonable profit on the investment can be made."

Commenting, Ford, Bacon and Davis pointed out that Lawrence and Wachter made two assumptions, both of which reflected favorably on possible development at Clairton/Edgar Thomson and adversely on Lorain, and whose validity they questioned. They criticized the presuppositions about depreciation and taxes on existing Clairton facilities and the choice of integrated transport costs on raw materials rather than tariff rates. The consultants concluded, "We believe it must be assumed that the development is to be permanent and all elements must be replaced from time to time whether they exist now or are a part of the new construction. Therefore the depreciation reserve should be set aside for the old as well as the new parts of the project." In short, they argued that estimates of both capital and production costs had been skewed in favor of Pittsburgh.[9] However, a few days before this report was made, the discussion was effectively foreclosed. Citing the urgency of taking action lest the USS share of national sheet capacity fall further from 13 percent to 10 percent a few years later, Ben Fairless, Carnegie-Illinois president, wrote, "We believe that a continuous strip sheet mill should be installed at Clairton immediately, with finishing facilities capable of producing 600,000 tons of sheet and tin plate annually." Puzzlingly, he added that the preference for Clairton was "primarily because of the important Eastern tin plate situation."[10] The letter helped clinch the argument, but the tale was to have one more episode.

On Friday, 15 January 1937, following the formal opening of a new 100-inch semicontinuous plate mill at Homestead, a dinner was held at the West Penn Hotel, Pittsburgh, for 600 business leaders and private citizens. In the course of the evening, William A. Irvin, now president of US Steel, and a former vice president of American Sheet and Tin Plate, announced the corporation would build 600,000 tons of strip mill capacity in the area. $15 million would be spent at Edgar Thomson, mainly on a one-million-ton slabbing mill, and $45 million on an 80-inch strip mill at Clairton works. It was indicated that at first 80 acres would be used at Clairton, and later another 67, but that the existing mills there would be retained. The new complex would be in operation in late spring or early summer 1938.[11] Within three months, there had been a radical change of plan. Instead of being on the already developed Clairton site, the new hot strip mill would be on a 600-acre virgin site at Dravosburg. This lay high above the Monongahela, along whose edge there ran the line of the Union Railroad, which would bring in the slabs from ET about six miles away. The new "Irvin Works" was expected to be in operation within the next year.[12] Ground was broken on 15 May 1937, and after 4.3 million cubic yards of material had been moved to make a plateau 227 feet above the river, construction got underway. The 80-inch hot strip mill, a 54-inch cold reduced mill for tinplate stock and an 84-inch cold reduced sheet and strip mill, altogether 51.5 acres under roofs in 17 buildings, were opened on 15 December 1938.[13] After considerable delay and by taking on extra site-preparation costs—and ignoring the prospect of higher overall production costs—Carnegie-Illinois had at last got a new strip mill to serve "eastern" markets. The whole project cost about twice as much as the 96-

inch strip mill with considerably larger capacity which J&L had completed in 1937, but Irvin would far outlive its more or less contemporary local rival. It should however be remembered that, through its acquisition of Otis Steel, J&L had acquired a strip mill in a prime location. The Republic and J&L mills in Cleveland and the Bethlehem Steel mill at Sparrows Point helped to "box in" the Pittsburgh mills to a smaller market area, and in 1938 an increase in the number of basing points made it costlier to "absorb freight," that is to sell into areas under other, nearer-basing points. In 1941 US Steel reported on the market area for Irvin. The account was matter of fact but echoed the limitations of the choice of location that the Carnegie-Illinois division had forced through: "Ohio and adjacent West Virginia counties, Pittsburgh and Philadelphia are major markets for its hot rolled sheets. Cold rolled sheets are principally shipped to Cleveland, other Ohio centers and Philadelphia. The Irvin works may also supplement Gary in the Detroit market with hot and cold rolled sheets in periods of peak demand, while Metropolitan New York is the major market for its large output of tin plate."[14]

While J&L and US Steel were planning Pittsburgh-area mills, three researchers of the economics of the industry recognized they had chosen less-than-optimum solutions and identified where the responsibility lay. The assessment was delivered in the measured words of an academic study: "One might be inclined to wonder, in the light of the present distribution of capacity in the industry relative to its markets, whether the erection of new mills in the Pittsburgh district would have been contemplated if the decision to erect new capacity had rested with a body interested in broad economic planning rather than with the executives of competitive units in the industry."[15] The remark revealed the New Deal ethos within which they were writing, but the implications for development were long-term, and the question they posed was an important one.

Most expansion in sheet mill capacity in the 1920s had been concentrated in smaller companies in the mill towns of Ohio and western Pennsylvania which had contained the bulk of this industry for many years. As one writer pointed out, in 1927 whereas the five main steel companies controlled 86.4 percent of structural shape capacity, 82.2 percent of that in rails and 73.9 percent in plate, for sheet and black plate their proportion was only 41.4 percent.[16] The strip mills which were added to the industry at the end of that decade and in the 1930s were either installed by some of these smaller companies at a relative few of the old-style sheet mill locations or, more commonly, built onto existing concentrations of integrated iron- and steelmaking capacity and rolling mills. There was one major exception to this pattern of growth. It involved the establishment and rapid expansion of the industry in southern Michigan. This was a response to the growth there of the nation's largest market for steel. In the mid-1930s, whereas the Pittsburgh area roughly consumed only one third of the steel it made, production of steel in the Detroit area was only one third as much as consumption.[17] When to this market attraction was added the advantages of a nodal position within the huge flows of iron and steelmaking materials on the Great Lakes, the area's lack until this time of a major part of the national steel capacity seems puzzling. An experienced English visitor to the United States in 1938 estimated that with a 50:50 percent pig iron–scrap ratio, burden assembly costs for making open-hearth ingots were about 47 cents more per ton in Youngstown than in Pittsburgh, which in turn was 19 cents a ton higher than in Cleveland or Buffalo. Costs in Chicago and Detroit were about 47 cents below Cleveland or Buffalo.[18]

From the beginning of the twentieth century, the Detroit area had contained a few merchant iron plants, and one or two rolling mills and small steelmaking operations. There was no fully integrated works. It was later reported that in 1907 the Great Lakes Engineering Company had interested J&L in a proposed location for such a plant along the Detroit River, but the combination of a Theodore Roosevelt blast against big business and the alternative and supposedly superior attractions of development at Aliquippa had killed off the idea.[19] As a result, during the first twenty years of the century, all growth of Lake Erie metal making was concentrated either around Cleveland or near Buffalo. By 1920 Michigan contained 16 plants making steel, but their combined annual capacity was only some 127,000 net tons; in short they were specialized operations, almost all of them foundries. About then American Sheet and Tin Plate (for whose blue annealed sheet plant at Gary, Michigan outlets were leading customers) planned a similar plant further east—location unspecified—to supply the automobile industry. The excessive cost of the projected plant caused its deferment several times, and it was in fact never built.[20] By 1922, the Detroit area contained no sheet capacity. In dramatic contrast, Cleveland then had four integrated works with a raw steel capacity totaling almost 2.25 million tons, and the Buffalo area three integrated works amounting to a capacity of 2.7 million tons. Yet in the first 20 years of the century, industrial growth around Detroit, as registered by the increase in the value added in manufacturing, was far greater than in any other of the main centers along Lake Erie, or in Pittsburgh. Through the 1920s a similar contrast in rates of growth did not occur, but by then manufacturing in Detroit was already second only to that in Chicago in the whole of the Great Lakes region, and most of this industry involved metalworking. The development of this market inevitably attracted important new units of the industry. Sheet and strip, leading lines of consumption in the state, became the focus of Detroit's new steel industry.

In 1930, the magazine *Mechanical Engineering* ran an editorial titled "Prospective Geography of the Steel Industry." This story pointed out how both production processes and marketing conditions were changing. The triumph of the open-hearth furnace meant that access to scrap supplies was of increased importance, though the writer exaggerated when he concluded, "Today the part played by the blast furnace in the manufacture of steel is gradually becoming of secondary importance, and the day may not be far off when all-scrap manufacture will become a predominant rule rather than the rare exception that it is today." He knew of one Detroit automobile firm which before World War I had sold all its scrap to a junkyard for $1,500 a year. "Nobody knows how many hundreds of thousands of dollars the junkman made out of this deal." By the mid-1920s, annual accumulations of scrap in the Detroit district were 700,000–800,000 tons. A few years later, when manufacture of the average car resulted in the creation of nearly 400 pounds of scrap, steelworks in the Buffalo area used 200–250,000 tons of scrap each year shipped in from Detroit.[21] Although generally of light grade, and therefore not the most desirable material for the open-hearth furnace, scrap produced in metalworking in southern Michigan could clearly be a major asset for any local steel manufacture. In overplaying its point on its future significance to the industry, *Mechanical Engineering* failed to acknowledge the area's other main advantage—that so much of the ore and coal used in American blast furnaces passed on lake carriers through Detroit. For iron manufacture, as well as for scrap-based steel production, it could be a highly competitive location. (In 1928, 36.5 million tons of ore moved along the Detroit River, and 700,000 tons of scrap were shipped out of the Detroit industrial area.) *Mechanical Engineering* recognized that if the changes it expected in steel manufacture occurred, increased importance would attach to the cost of carrying finished products to market, and as this amounted to about $1 per ton for a haul of 100 miles, proximity to centers of major consumption would become more advantageous.[22]

In June 1923 the Federal Trade Commission, in the course of its enquiry into Pittsburgh Plus pricing, received estimates of steel consumption from four Michigan automobile manufacturers. By 1928 the nation's automobile industry was estimated to consume 6.3 million tons of steel of all types, of which the Detroit district accounted for about 90 percent. Next year, when Michigan ranked fourth among the states in steel consumption, it was 17th in production. It was then reckoned that freight charges on finished steel from Pittsburgh to Detroit were $4/ton higher than the switching charges a Detroit producer would have to pay to deliver his steel to local auto plants.[23] Clearly there were many incentives to make good the shortfall of local output relative to local consumption. By 1930, two integrated works had been built, and steel capacity reached 1.26 million net tons. Advance in sheet and strip production played a central role in their development.[24]

During its early years, and again in the early 1930s, steel expansion in Detroit was largely homegrown; in between there were important investments from outside. The pioneer was the Ford Motor Company, then dominant in the automobile industry. In 1915, in order to expand its production, Ford purchased large tracts of land along the River Rouge in Dearborn. Work started on an iron and steel plant there in 1917, but it was stopped for a time by wartime pressures. In September 1918, construction of two blast furnaces began, and the first coke ovens were at work 13 months later. Electric furnace steel was produced in May 1923. During the same year Henry Ford announced his company would install open-hearth furnaces, a blooming mill and rolling mills for bars and narrow strip. Even so, for a time Ford seemed so little committed to making River Rouge into a major operation as to be reported as planning an integrated iron and steel mill close to its new automobile construction works near St. Paul, Minnesota.[25] Though it now produced large tonnages of finished steel, the company continued to buy from outside. Other motor firms remained wholly dependent on purchases. There seems to have been a tacit understanding among the major steel companies that no one of them would attempt to jump ahead of the others by building in the Detroit area. Details of that reported agreement are not known, but a consequence was that the way was open for smaller firms and newcomers.

Given its rapid increase as a consumer of steel, it is understandable that three of the ventures which began in Detroit in the 1920s stemmed from their promoter's experiences in marketing. In the early 1920s one salesman, George R. Fink, then in his mid-30s, came to the area as representative of the West Penn Steel Company of Pittsburgh, a company whose sheet was highly regarded. Realizing the possibilities for a large sheet steel plant in the area, Fink tried to interest his employers. He failed in this attempt, and decided to build the mills himself. In 1922 he formed the Michigan Steel Corporation and invested in Ecorse, some 10 miles from downtown Detroit. The annual capacity of 40,000 tons of sheet in this plant was almost completely devoted to auto body, hood and fender material. With natural advantages and aggressive management Ecorse did well. By the end of the decade it could make upward of 100,000 tons of sheet a year, from purchased sheet bars.[26]

The origins of another steel concern were almost contemporaneous with Fink's. In 1922 two brothers, Wendlin and Arthur Schrage, and a third man, Clyde Craine, organized a new business as representatives and warehouse operators for outside steel mills. They realized how difficult it was to deliver on time the types of steel on which their customers depended. This applied especially to cold-rolled strip. Responding to this situation, the partners pooled their resources and set up the Detroit Steel Corporation with a capitalization of $500,000 to build a small cold mill to supply automobile, hardware and typewriter manufacturers.[27] Whereas Michigan Steel would become the basis for a much bigger, local steel and strip enterprise, Detroit Steel over a longer period helped shape steel and strip mill development well away from southern Michigan. Another, later project had a wider historical interest. In 1934 the Rotary Electric Steel Company, established in Detroit three years before, began to build a cold rolled strip mill. The organizers of Rotary had been Naugle and Townsend, of the Butler hot strip mill.[28]

Yet another project was homegrown. After Donald B. McLouth graduated from college in Ypsilanti, he worked in an automobile plant. Finding sizeable pieces of steel sheet were discarded as scrap, he decided to enter the scrap business, organized a company, bought this material and for a time sold it on to makers of smaller sheet-using products. In 1934 he formed McLouth Steel Corporation, bought a 20-inch Steckel reversing, one-stand rolling mill from the Cold Metal Process Company of Youngstown and began to produce hot rolled strip at Trenton, on the south side of the Detroit metropolitan area. At this time McLouth could roll strip of up to 12-inch width, and had a monthly capacity of 4,000 tons. Overheads were extremely low, the capital cost being $5–$6/ton of annual capacity, compared with many times that for a strip mill.[29] It was a modest start, but eventually from origins in brokerage and steel jobbing, McLouth would move on to bigger business.

One other important mill project came from outside the Detroit area. Its history was particularly interesting, for it showed how limited the opportunities were now becoming even for efficiently designed old-style sheet mills in a prime location. In 1919 the Newton Steel Company had been incorporated with a capital of $3 million. It erected sheet mills almost 20 miles to the northwest of Youngstown at Newton Falls.[30] By 1927 Newton had 20 mills producing full-finished sheets, cold rolling mills, and had done so well that Pittsburgh and New York bankers and engineers felt justified in offering $42.50 a share for its common stock. In addition to $4.25 million capital, it had another $1.25 million in preferred stock. It was rumored that both Ford and the Fisher division of General Motors made moves to acquire control.[31] By early 1930, Newton assets were $13 million, and it was completing a new sheet mill in the Detroit area, helped in part by an injection of capital from Fisher Body. A 600-acre site was acquired on the Raisin River two miles east of Monroe. Ground was broken in early spring 1929.[32] Even though there were serious construction problems due to the marshy site, federal government engineers were already dredging the two miles of river through to deepwater on Lake Erie, and it was expected that sheet bars bought on the open market could be delivered direct to the works. In 1929, the company spent

$540,000 million on its Newton Falls mills, but $7.0 million at Monroe. By January 1930 eight new sheet mills, with a monthly capacity put at 20,000 tons of auto-body stock, were ready for production. They were high-capacity mills with a tonnage equaling that of the more numerous mills at Newton Falls. In 1930, plans for Monroe were modified to provide for a possible full-finished sheet capacity of as much as quarter of a million tons, and a five-year program of further developments was reported.[33] Unfortunately, it proved too late for such improvements and initiatives, partly because strip mills were by now too firmly established, and in part as demand fell away as the Depression deepened. By late summer 1930, Monroe was periodically idle. Net profits of Newton Steel had been $799,000 in the June quarter of 1929 but were only $41,000 a year later. In the September quarter of 1930 there was a large loss.[34]

By the late 1920s rumors were rife of further developments in southern Michigan. They were encouraged by the suggestion that if the big steel firms failed to act, other automobile makers might follow Ford into making their own steel. In autumn 1928, it was said that General Motors was planning blast furnaces and a steelworks in the Saginaw district, and there was talk of a similar plan for Detroit. One Ohio steelmaker was reported as likely to come in. By mid-1929, there were said to be plans for blast furnaces, open-hearth furnaces, and bar and strip mills for the Monroe site of Newton Steel. Some pressure seems to have been brought to bear on US Steel for a Detroit area plant, but it was recognized that it was involved in such ambitious plans for Chicago that Detroit hopes were dampened. Within two months, Armco had optioned a 200-acre site at Wyandotte. There were also rumors that an eastern company was contemplating a mill near Huron between Toledo and Detroit.[35] None of these schemes materialized; some of them may have been the product of little more than speculation. The most important step was taken by a much smaller enterprise, but one which had already succeeded spectacularly.

After six years of aggressive business in a generally expanding economy, the Michigan Steel Corporation had a 1929 sheet capacity of about 125,000 tons. In February 1929, it was announced that Fink "contemplated" building a completely new works, equipped with open-hearth furnaces to produce structural shapes, light plate and "hot-rolled flats." The scheme evolved rapidly, and by early March plans were revealed for a 400,000-ton steelworks costing $20 million. It would work closely with the Hanna Furnace Company, which had two blast furnaces only a mile away. Some large automobile manufacturers and other important consumers were among its stockholders. The project was so enthusiastically welcomed in Detroit that within three weeks Fink received offers of 50 percent more capital than he required. Construction began in May in a large swampland area alongside the Detroit River in Ecorse. It was said to be the only site of sufficient size within the Detroit switching area with suitable rail and water connections. 75,000 wooden piles to be capped with concrete had to be driven down to the rock foundation, 1.6 million tons of sand was hydraulically pumped in from boats, and dry fill was brought in by truck and rail. In fall 1929, as the work went on, the new "Great Lakes Steel Corporation" merged with MA Hanna, which, in addition to two local furnaces, had six blast furnaces elsewhere; controlled coal lands in Pennsylvania, Virginia, and Ohio; owned large ore properties in the upper Lakes; and operated a fleet of ten ore carriers. Hanna ore-mining operations also supplied Weirton Steel, and on October 1, 1929, a new holding company, the National Steel Corporation, was organized to include Great Lakes, Weirton and Hanna.[36]

The flat, virgin site made it possible to pay careful attention to plant layout so as to minimize backtracking of materials. The first Ecorse open-hearth furnace was tapped in August 1930, and early the next month steel was rolled on its 34-inch strip mill. Soon this mill was modified so as to roll 38-inch widths, and before the end of the year, Great Lakes Steel had acquired the assets of Michigan Steel Corporation, providing it with up to 250,000 tons of sheet capacity including full-finished material up to 72 inches in width. In turn, Ecorse supplied sheet bar and strip mill breakdowns to what had essentially been its parent enterprise. By August 1931, two years after starting, Great Lakes completed its first program of development at Ecorse at a cost of \$29 million.[37] Even in these exceptionally difficult times, National Steel was an immediate commercial success, in the 1932 depth of the Depression making a modest profit of \$1.6 million when Bethlehem Steel lost \$19 million and United States Steel \$71 million. Within four years it had doubled its number of open-hearth furnaces; in five years a second, much wider strip mill was installed. Before this, the initiative for a time passed back to the pioneer of big steel projects in the area, now very much second in terms of capacity. Early in 1934, Ford announced expenditure of \$10 million to build a hot strip mill. This was commissioned in the second half of 1935.

By early 1936, the Detroit area had three hot strip mills, a concentration of capacity exceeded only by the four mills of the Chicago area. At that time there were no strip mills in the immediate vicinity of Pittsburgh. Commenting on the fact that National Steel had been extremely profitable and had operated at up to 100 percent through part of the Depression, *Fortune* again listed the exceptional advantages of the Detroit area and looked ahead to investment there by rivals: "That there will be others is beyond question; no more logical location could be imagined, with cheapness of raw material transport added to nearness to the auto industry and low scrap prices."[38] Why, despite frequent rumors, did the bigger, established steel companies still fail to come into the prime Detroit market? A mixture of considerations seems to have been involved. One was an apparent lack of suitable sites. *Fortune* reported in 1932 that plant engineers from "many" companies had previously visited Detroit and reported against the location because of unfavorable physical conditions.[39] Yet there is also evidence that during and after the Depression outsiders did attempt to break in. Sometime in the first half of the 1930s, a "prominent" steel company approached Henry Ford with an offer to purchase his Dearborn mills, and when this met with no encouragement, made clear that if it could not acquire existing properties, it would build its own flat-rolled product mills there. Two other major companies were reported to have had options on mill sites at the southern end of Detroit.[40] That, despite this, no major steel company broke into the district points to a third and perhaps vital consideration: no outsider could risk venturing into Detroit, because if it did so, its main rivals would feel compelled to follow. There was in effect a "gentlemen's agreement" that no one of them would take the lead.[41] The deterrent effect of this survived even when, fearing that the motor firms might build their own mills, the steel industry adopted "Resolution 21," which granted to the automobile industry price concessions by "absorbing" freight charges. Under this arrangement, delivered prices to consumers in Detroit were fixed below those in, for instance, Toledo, through which much of the supply from Cleveland or Pittsburgh had to pass. As a result, in the mid-1930s Michigan consumers could obtain steel for $3–$5/ton less than the basing point plus rail freight charge.[42] After the abolition of most price differentials between basing points in 1938, freight charges, which Pittsburgh mills had to absorb in making shipments to Detroit consumers, increased from $1.80/ton to $3.80/ton.[43] Even now, however, the imbalance between demand and supply in southern Michigan remained. In 1938, the estimated capacity of the strip mills in the Pittsburgh area was 1,500,000 tons and consumption of their products only about 110,000 tons; in the Detroit area capacity was 1,400,000 tons and consumption 1,650,000 tons.[44] This pattern was to persist. The continuing mutual suspicion among the older, bigger steel firms was brought out in a letter written in January 1936 By Charles Hook of Armco to the then president of the American Iron and Steel Institute, Eugene Grace of Bethlehem Steel. Hook urged that steelmakers should resist the pressure from the automobile industry to reduce prices. A price was needed which gave steelmakers a reasonable return on their huge investments, but, as he pointed out, successful resistance was only possible if the steel companies treated all customers equally and did not resort to undercover methods.[45] As things turned out, no major strip mill operator from outside would ever build in Detroit or the southern Michigan industrial area.

The growth of the Detroit sheet and strip mill industry was an outstanding example of the new, looser controls on the location of the steelmaking. That it was so long delayed was a witness to the strong force of inertia even in the world's most dynamic industrial economy. Even when one firm overcame this, the absence of the older majors showed that strong institutional controls could frustrate economic logic. Nonetheless, although absent from Detroit, the longtime leaders in steel had to make their own, costly and sometimes painful responses to the new product structure of their industry. They had missed the pioneering days in the 1920s; now they made their way in the midst of depression or during the slow, uncertain revival of the mid and later 1930s.

NOTES

1. G. C. Kimball in *AISI Yearbook* 1935, 162, 163.
2. *IA* 19 December 1935, 44, 16 December 1937, 49.
3. FB&D *Report 100* "Organization," August 1935.
4. *IA* 17 March 1932, 687; Ford, Bacon and Davis *Report 7*. Study of AS&T and Carnegie-Illinois Steel with respect to a possible merger, 1 April 1936; *Fortune* March 1936, 204.
5. *BFSP* November 1928, 1473.
6. Ford, Bacon and Davis *Report 150,* Study of Carnegie-Illinois Steel Corporation, 26 November 1937.
7. Ford, Bacon and Davis *Report 8,* 1 April 1936.
8. Ford, Bacon and Davis *Report 11,* 12 June 1936; report 200, vol. 2 Nov 1937.
9. Ford, Bacon and Davis *Report 34*, "Relative merits of Clairton, Pa and Lorain, Ohio for the proposed sheet mill development," 15 December 1936.
10. Letter of Ben Fairless, 11 December 1936; Ford, Bacon and Davis *Report 36*, "Final recommendations respecting sheet and tinplate program at Pittsburgh."
11. *IA* 21 January 1937, 58.
12. *IA* 25 March 1937, 70.
13. *Mechanical Engineering*, February 1939, 156.
14. *TNEC monograph 42,* "The Basing Point Problem" 1941; Paper "Some factors in the pricing of steel" especially 17; *Economics of Iron and Steel Transportation,* 170–172.
15. Op cit, 340.
16. McCallum 1931, 131.
17. *Fortune* April 1936, 130.
18. Russell 1938.
19. *ITR* 14 March 1929, 757.
20. *IA* 21 June 1923, Account of Pittsburgh Plus hearings.
21. *AISI Yearbook* 1924, 409, 410; *IA* 3 July 1930.
22. *Mechanical Engineering* 52, No. 9, September 1930, 869; *ITR* 18 April 1929, 1066
23. *IA* 8 August 1929, 351.

24. *IA* June 1923, 1859, 8 August 1929.

25. *IA* 6 May 1920, 1276, 1 January 1925, 52; *BFSP* May 1927, 212, 214, 215, August 1951, 927–30.

26. *IA* 3 January 1923, 50; *ITR* 18 April 1929, 1066, 18 July 1929, 173; *IA* 8 August 1929, 351.

27. *IA* 3 May 1923, 1284; see also useful account in "Steel: A History," quoted *YS&T Bulletin* January–February 1968, 12.

28. *IA* 5 July 1934.

29. *BFSP* June 1934, 345; on the Steckel mill, see *IA* 19 May 1949, 860.

30. *BFSP* August 1919, 409.

31. *IA* 27 October 1927, 1196.

32. *IA* 27 October 1927, 6 February 1930, 478; *ITR* 28 February 1929; *Steel* 18 April 1929, 1066.

33. *IA* 2 January 1930, 102, 10 April 1930, 1064, 29 May 1930, 1616; *ITR* 18 April 1929, 1066, 23 May 1929, 1433, 1434.

34. *IA* 18 September 1930, 815.

35. *ITR* 18 April 1929, 1066, 23 May 1929, 1434, 18 July 1929, 173,174, 19 July 1929, 734, 19 September 1934, 734.

36. *IA* 28 February 1929, 622, 7 March 1929, 683; J. A. Clauss "The Development of the Great Lakes Steel Corporation" *Iron & Steel Engineer,* September 1936; *Fortune* May 1931, 87, 89, June 1932, 31–37, 86–89.

37. *IA* 27 November 1930, 1651, 13 August 1931, 449.

38. *Fortune* December 1935, 142.

39. *Fortune* June 1932, 35.

40. *IA* 5 July 1934.

41. *Fortune* June 1932, 36; Stocking 1954, 132.

42. Burn 1936, 313,314.

43. *Pittsburgh Business Review* 30 September 1938, 2.

44. *IA* 8 February 1940, 78; TNEC *Monograph 42*, "The Basing Point method of quoting delivered prices in the steel industry," 85.

45. C. R. Hook to E. G. Grace quoted TNEC *Monograph 42*, 95.

Chapter Ten

Hand Mills, Continuous Mills, and Social Problems

"When the pace of technological change is rapid, the business enterprises which grasp the new opportunities for gain bring to pass mass changes in economic conditions, and unwittingly produce a host of economic problems. All of these problems may be summed up in the question: How can society improve its economic organization so as to make full use of the possibilities held out by the march of science, invention, and engineering skill, without victimizing many of its workers."[1] In that brief statement, published in 1933, the authors of a study of *Recent Social Trends in the United States* recognised that technological progress, such as that eminently illustrated by the continuous hot strip mill, involved major economic and social costs. The steel companies had to pay the first; the second was borne by steelworkers.

Investment in strip mills required a huge outlay on the part of the steel companies and rendered a large part the earlier investments by them and by their competitors in old-style mills either wholly wasted or largely redundant. At the end of the 1930s, Frank Purnell of Youngstown Sheet and Tube made an effort to summarize this aspect of the technical revolution. Generalizing, he reckoned that in 1926 there had been 1,320 stands of hand mill rolls for sheet, strip and black plate. Depreciation schedules had been set up on the basis of approximately $100,000 per stand, and amounting to about $132 million for the industry. Since then 28 continuous high speed mills had been built at a cost of some $500 million.[2] The financial problems of the industry revealed by such figures were real, but even more obvious and painful was the situation of the many thousands of often skilled workingmen who were thrown out of employment by the massive commitment to the radically new methods of production. That there had been tremendous material progress was undeniable, but its human cost was high.

On Wednesday, October 19, 1938, schools, stores, and some industrial plants were closed for the day in Ashland, Catlettsburg and neighboring communities in Kentucky and across the river in Ohio as the population celebrated the 15th anniversary of the continuous sheet mill. George Verity and Charles Hook were scheduled to speak. Photographs were taken of John Tytus being feted.[3] Not all sheet-, tinplate- or strip-producing communities shared the enthusiasm demonstrated that day in and around Ashland. Only a few months before, Charles Hook had remarked that there was no need to fear overcapacity, because the continuous mill by improving uniformity, quality and the size range of the product and at the same time lowering its costs and prices had greatly extended old markets and made new ones. However he added that this was "in normal times."[4] Through most of the 1930s, times had been far from "normal," and while the strip mill industry expanded, old-style mills suffered severely, many companies or plants failed and workforces and sometimes whole communities were plunged into deep depression.

It was clear from the early days of the continuous sheet and strip mills that at high levels of utilization, their unit costs of production would be lower than those in hand mills. A 1927 assessment and projection was that they might save $15/ton over sheet mills, that seven continuous mills "could take care of all the sheet rolled today in the United States," but that smaller orders and odd sizes "will be always left to the present non-continuous mill." The same account suggested that, as the Butler mill had been an outgrowth of the old-style strip mill rather than of sheet mills, it was usually strip mill and not sheet mill labor which was being taken over to run the new strip mills. Time would in fact show that it was the more numerous sheet companies and the mill towns to which their operations had given rise which suffered most.[5]

During the second half of the 1920s, as investment in strip mills began, the rapid increase in the consumption of sheet meant there was room for both new and older processes of manufacture. After 1929 that was no longer the case. Output of sheet, strip and black plate fell during each of the next three years, and 1932 sheet and strip consumption by the automobile industry was less than one third the 1929 level.[6] Consumption and production of thin flat-rolled products revived after that but did not again exceed the 1929 figure until 1936. Notwithstanding this most severe of crises, output by the continuous mills did not decline in a single year. By 1936 the tonnage they produced was not far short of six times the 1929 level, but their capacity was about 18 percent greater than their record output that year. A corollary of this sort of expansion, of the economics of the new process and of the keenly competitive conditions, was that prices declined. *Fortune* summed up the situation by 1931—the "normal" cost of 24 g sheet was $60.65/ton, but the strip mills had cut about $5 off this and yet had also made better sheet.[7] Between 1923 and

1937, the price of automobile fender sheet fell from $135/ton to $68/ton—and its quality was improved. In 1934, one large maker of sheets received less than 54 percent the price he had been paid in 1923. The 1936 prices for hot rolled sheets and hot rolled strip were, respectively, 81.0 percent and 83.0 percent of their 1926 levels. Three years later, Eugene Grace reckoned that the continuous mills had cut the prices for sheet and strip by $6–$8/ton. Such reductions meant little or nothing in terms of the price of most end products in the finished domestic durable industries—a refrigerator or a car, etc.—but in an oversupplied market consumers of sheet and strip could press for and usually receive price concessions. After their extremely high initial capital outlay had been made, strip mills were extraordinarily productive units, at least when operated at high capacity. Naturally, therefore, their owners aimed for a high utilization rate. The old-style sheet mills paid the price. By the mid-1930s, it was reckoned that a 400,000-ton strip mill had a productive capacity equal to 40–50 old-style mills; 126 men in strip mills (ignoring the labor force in ancillary activities such as iron- and steelmaking and in primary rolling mills) could produce as much tonnage as 4,512 men in hand mills. In 1929, 1,400 old-style mills had an annual capacity of about 8.4 million net tons. Over the years 1926–1937, 27 continuous sheet and wide strip mills were installed. Their capacity was put at some 14.7 million net tons.[8] Even in the good year 1937, total production of thin flat-rolled steel was below 13 million tons. In short, in tonnage at least there was now little or no need for the old-style mills. The inevitable outcome was a sharp though irregular fall in their

Table 10.1. Production of sheet, strip, and black plate by continuous mills and output left to old-style mills, 1925–1937 (thousand net tons)

	Total	*Cont. mills*	*Older mills*
1925	6,780	181	6,599
1926	7,087	200	6,887
1927	6,615	512	6,103
1928	7,943	814	7,129
1929	8,303	1,380	6 ,923
1930	6,595	1,403	5,192
1931	5,055	1,522	3,533
1932	3,440	1,572	1,868
1933	5,663	2,619	3,044
1934	5,530	3,168	2,362
1935	8,104	4,850	3,254
1936	10,780	c.8,000	c.2,780
1937	12,088	c.10,000	c.2,088

Based on Eppelsheimer 1938, p. 195

output. The dimensions of the problem were indicated in figures quoted by D. Eppelsheimer of Middletown, Ohio, in a paper published in 1938.[9]

The hand mills did not give in without a long fight. There was much to be lost. As late as 1931 R. J. Wean put the capital invested in the 700 or so old-style sheet mills at $175 million. He believed that with continuous heating of the sheet pairs and use of mechanical feeding and catching, conversion costs in hand mills could be reduced to only about 61 percent of those in conventional practice.[10] But that year output of hand mills was little more than half the level of three years before, and such claims were beginning to sound hollow. A year later, as the trade reached its lowest point, some mill operators still managed to remain optimistic.[11] One commentator was much more circumspect. In the rolling mill industry generally, 1931 had not been spectacular. In place of "the glory of an offensive campaign for great tonnages and large profits," there had been "a struggle for existence, a determined and grim defensive." The writer was guarded in his expectations, though granting that all the improvements being made by the hand mill industry might "doubtless defer for some time" the advance of the new technology.[12] Five years later, most of the smaller producers still existed, but they were increasingly specializing—on small orders or on unusual sizes or grades of sheet. Their coexistence with the strip mills was an uneasy relationship in which either one side or the other might invade at least the fringes of each other's territory: "There are two different spheres of activity for the continuous mills and the smaller company and yet each can without undue harm reach over into the other's field."[13]

In spring 1931, an assessment from within the industry put national sheet and hot rolled strip/sheet capacity at more than 12.5 million tons—5.89 million tons for sheet mills, 0.81 million for jobbing mills, 1.33 in continuous sheet mills and 4.75 million tons in hot strip mills.[14] By mid-1936, capacity was much greater—8.68 million net tons in old type mills and 9.46 for continuous strip mills—yet according to the same source, 1935 production was no more than 7.18 million tons.[15] Although by the last quarter of 1936 demand had revived so much that many old hand mills were again in full work, in the following year major companies were spending lavishly on strip mills, and hand mills were being closed. Three years later there was another surge in activity in reopened old-type mills, but this too proved a passing phase.[16] So too were attempts by old producers to compromise with the new technology, as when in 1940 the Follansbee Steel Corporation built a new cold reduction plant on which to roll down purchased cold rolled strip.[17] It was all too late. Conclusive proof of this was provided in the early 1940s in a comparison of the man-hours per ton of product on the new US Steel Irvin hot strip mill and in the old-type hot mills at the Corporation's large Shenango works—5.22 man-hours as compared with 24.91.[18] In the early aftermath of the Butler mill, it had seemed possible that the strip mills and old style mills could coexist, if the strip mills supplied mill "breakdowns" to be finished on the hand mills. R. J. Wean had seen this as a way to minimize loss: "Surely the development of this market is worth considerable thought. The cooperation of the wide strip producer and the sheet steel producer would not only work out to the economic advantage of both, but to the entire steel industry as well, and most certainly to the stockholders who have invested their money in these companies."[19]

Discussion about the impact on the sheet companies usually failed to focus on the most critical sector of all, their workforces. Starting in the late 1920s, but proceeding much further through the 1930s, many thousands of mill workers lost their jobs. In the nature of sheet-making operations they were often highly skilled, well-paid men, part of the aristocracy of steel industry labor. As a leading trade journal had recognized as early as summer 1927, "The ordinary sheet mill is probably one of the largest employers of labor in the entire industry, and, what is more, the labor used in the sheet mill is of a highly skilled, high-priced kind, with an organization unusually well able to enforce its demands."[20] These men were now humbled, though their downfall would be spread over many years.

One example of the impact of rundown and consequent technological unemployment was provided by the experience of Empire Steel. Formed in 1927, it had at the start ranked as the nation's fifth-largest sheet company. Over the next two years it borrowed heavily from the banks, and when depressed conditions arrived was unable to service these loans through earnings. By late summer 1930, though having a nominal sheet capacity of 400,000 tons, its mills in Cleveland and at Ashtabula had been shut down for some time and three works in Niles were operating only intermittently. The Empire steel plant at Mansfield obtained its pig iron from the Toledo blast furnaces of Pickands Mather, and this iron maker now gained control of the sheet group. Early in the following year, Empire Steel capital was reduced from $14.2 million to $7.6 million. Soon it was in receivers' hands. It continued to operate only three of its works, mostly on automobile sheet—and reportedly at prices some bigger companies were not disposed to meet.[21] Some years later the Steel Workers Organizing Committee estimated that Empire laid off 1,950 men in Ashtabula and Niles alone in the three years 1929–1931. In 1937 and 1938 a further 735 jobs were lost, this time in Mansfield.[22]

Ill-fortune for some provided opportunity for others. By 1935 the Falcon plant had been bought by a group headed by H. A. Roemer of Sharon Steel. It was renamed the Niles Rolling Mill Company, and there were plans for remodeling it so as to make 60,000 tons of sheet a year. As the Niles Rolling Mill Division of Sharon Steel, it survived 20 years more, by which time annual sheet mill capacity was 150,000 tons. Continental Steel made smaller displacements, but 450 of its jobs were lost in Canton, Ohio, and in Indianapolis and Kokomo, Indiana. In 1940, it started a new automatic sheet mill at the last of these works.[23]

The major steel companies which installed strip mills also had to tackle the problem of redundancy in their old-style mills. In late summer 1939, Youngstown Sheet and Tube commissioned a 54-inch hot strip mill at Indiana Harbor. Improving the quality of the output required not only a strip mill but also investment in new open-hearth furnaces, a blooming mill and so on. Twenty-four old-type hot mills were closed and 1,200 hand mill jobs were lost.[24]

The steel companies argued that strip mills not only represented technological progress but also brought a major advance in working conditions. Most of this was incontrovertible, but representatives of the men emphasized the high social cost involved with this gain. In 1941 Armco contributed a chapter on the iron and steel industry to a book: *The Development of American Industries: Their Economic Significance.* In it, old-style sheet mills and continuous strip mills were compared. The account of the former contained a number of words evoking intense physical labor—"grasping," "passing," "dragging" and "pulling the sheets apart." Summarizing the "disadvantages of the manual-type mill," the authors wrote that "a strong back is essential to the sheet-mill worker on the old manual-type mill. Moreover he must be a skilled artisan, for each sheet is rolled according to specifications dictated by hand and eye. Of course, a high degree of uniformity cannot result, because the hand and eye are not as reliable as the machine." The authors quoted John D. Knox of American Sheet and Tin Plate on the benefits of working in a strip mill: "Seldom has a new departure rendered better service to mankind than the continuous mill . . . all this actually accomplished mechanically. Roller tables take the place of tongs in the manipulation of steel into the rolls; automatically controlled furnaces take the place of . . . heaters. No heat-soaked floors to fatigue the mill crew, and the discerning eye perhaps will notice other things about the mill that signify the reign of a new order and emphasize that new values have taken the place of the old."[25] No doubt all of this was true, but the cost of the progress was not always acknowledged. Fortunately, the first chapter of the same book quoted William Green, president of the American Federation of Labor, and provided a balance: "In the steel industry between 1929 and 1939, 53 old-process mills producing plate, tin plate and sheet were permanently abandoned and replaced by the continuous-strip process. This change has displaced 38,470 workers. In addition, in large integrated mills that are planning to install the continuous-strip process, 22,950 workers will be displaced if and when the new process comes into operation.

The strip process in these large companies put a number of small companies out of business, throwing out an additional 23,350 workers. The total of those already displaced and those who will be displaced if present plans become effective is 84,770."[26] Green's figures had a doubtful precision, but the scale of the redundancy which was enforced by this technological change was unquestionably huge.

Those who took up the cause of displaced workers sometimes took their opposition to new technology to irrational lengths. An example was a proposal made to a conference of 200 delegates at an early 1934 meeting of steel unions. It was noted that 400–600 tons a day could be rolled by the same number of men in a strip mill as could make only 25 tons on a hand mill, and that 1,600 men in the 12 continuous mills then existing could turn out as much as the 32,000 in the 636 hand-operated mills, many of which were already idle. It was suggested that the Government should spend $150 million to buy up and scrap all the continuous mills.[27]

At the end of the 1930s evidence of the human impact of the strip mill was given to TNEC by the Steel Workers Organizing Committee. The devastating effect was summed up by John D. Knox, who though earlier quoted on improvements in working conditions reported another aspect from a conversation he had had with an official operating a strip mill. Knox had asked about technological unemployment. The other man talked of the labor requirements for 24-hour operations: "With 126 men in the slab yard, on the furnaces and overseeing the mill you can turn out 2,500 tons a day whereas it would take 96 sheet mills of the conventional type with a combined crew of 4,512 men to produce an equivalent tonnage." The resulting closure of hand mills caused large-scale redundancy and hardship. Sometimes the companies handled this with extraordinary insensitivity. For example, on March 29, 1940, Republic Steel sent the following notice to 500 workers in Massillon: "We regret to advise you that on account of the permanent discontinuance of operations of the Massillon sheet mills your services are hereby terminated. Please find enclosed your copy of the 'Termination Notice to Employment Office.' This form should be presented to the paymaster to secure any earnings which may be due to you. Also find enclosed 'Workers Copy' of form UC 406 'Separation Report for Total Unemployment' as provided under Unemployment Compensation. (Signed) Yours very truly, Republic Steel Corporation."[28]

The biggest program of plant closure was that carried through by US Steel. The ten years during which, as chairman of the finance committee and then of the Corporation, Myron Taylor was a decisive force in decision making coincided with the advance of the strip mill. When he retired in 1938, he presented stockholders with a comprehensive review, "Ten Years of Steel." This recognized the greatest change in technology in his period of office had been "the demand for a character of finished steel that could only be made on continuous mills with highly refined processes." The implications for older processes of manufacture had been serious but unavoidable, for continuous mills "represent large capital outlays and, while they are very economical if running full, they are very costly if running only on part time. Thus the concentration of production became a factor of such importance as sometimes to outweigh freight rates. In certain of our plants the obsolescence, taking that word in its broader significance, was such that their abandonment was scarcely a matter for discussion."[29] What may have been "scarcely a matter of discussion" for management, can rarely have been regarded as such by labor. Myron Taylor listed the plants abandoned during these years. They involved various fields of production, but the Sheet and Tin Division, formerly American Sheet and Tin Plate, was prominent. Stockholders were told how much steelmaking and rolling mill tonnage had been abandoned, dismantled or sold, and that $211 million had been written off on account of retirements and replacements, but Myron Taylor was not so explicit about jobs lost. Giving a rather different list for the ten hand mill works the Carnegie-Illinois Steel Corporation closed between 1931 and 1937, the research department of SWOC came up with a loss of 10,450 jobs.

Particularly in the case of the hand mill companies, much of the displacement of labor was concentrated in small towns, which until then had often been growing. Loss of their local mill had a sudden impact on the economy and on social well-being. In the forty miles along the Ohio between Wheeling, West Virginia, and Toronto, Ohio, the jobs lost from hand mills in the four years 1933 through 1937 amounted to 4,950. Even for bigger, integrated firms, technological unemployment had become a serious problem.

Table 10.2. Old-style mills closed by the United States Steel Corporation, 1928–1937

1928–1930	Old Meadow Works, Scottdale, Pennsylvania; Pittsburgh Works, New Kensington, Pennsylvania
1931	Pennsylvania Works, New Kensington, Pennsylvania; Aetna-Standard Works, Bridgeport, Ohio; Chester Works, Chester, West Virginia; Leechburg Works, Leechburg, Pennsylvania
1934	Wellsville Works, Wellsville, Ohio; New Philadelphia Works, New Philadelphia, Ohio
1935	Dover Works, Dover, Ohio
1937	Cambridge Works, Cambridge, Ohio; Sabraton Works, Morgantown, West Virginia; Scottdale Works, Scottdale, Pennsylvania

Source: US Steel, "Ten Years," 1938

In December 1936, Philip Murray's advisor and research director in SWOC wrote to its staff representatives in steel centers to ask for their cooperation in estimating the number of workers who had been or would be thrown out of work by strip mills already built or under construction. Among the many letters he received were records of major shrinkages in workforces and of the consequences for individuals, their families and hometowns or districts. The field director in Cleveland replied promptly, highlighting the experience of workers at Otis Steel Company. Otis had operated six hand mills which had together made about 270 tons of sheet a day; in 1931 it installed a strip mill which produced up to 100 tons of strip sheet an hour. Three hand mills were torn down, and the remaining three averaged less than three days of work a week. He compared other aspects of the situation. Old-type mill workers "are eased out one after another. At the most they only get an average of about two or three days per week . . . while in the new continuous mill they hire young men just out of school, because it is more or less a 'button pushing' job. . . . These kids are only paid a little above laborer's wages."[30]

Bethlehem Steel followed a similar course. The company took over the Seneca Iron and Steel Company in the early 1930s. If fully at work, this company's Blasdell mills employed about 1,800. At the beginning of 1936, Bethlehem commissioned its first strip mill at the nearby Lackawanna works. For a time there seemed no adverse impact on Blasdell, but during the year it was learned that as compared with its current complement of 20 roughing and finishing mills only six finishing mills would survive, rolling down material already roughed on the strip mill. By early 1939, the Seneca East plant had been closed, and the West plant was to be converted into a galvanizing mill with jobs for only about 300. The labor representative recorded, "When the men were laid off they were promised by the company that they would be absorbed in other departments of the plant, but very few men were ever placed."[31] At US Steel, the vice-president for industrial relations, Arthur H. Young, wrote to William Irvin in June 1936 to object to a suggestion from the Corporation's consultants that men with less than five years of service should be included in a scheme to give at least one week's notice or pay in lieu thereof to dismissed employees.[32]

Both employers and workers, as well as those who were trying in this period to organize the latter into unions, struggled to find a remedy for localized unemployment. US Steel at the beginning of 1937 had 11 black sheet works with a total capacity of 1,299,000 tons and 4 strip mills of 2,131,000 ton capacity, the latter in two locations, mainly at Gary and the other in the McDonald works, near Youngstown. In 15 sheet, strip mill and associated works in the Steubenville district of SWOC in 1936, 32,050 men were estimated to be at work. By December 1938—a year of sharp recession—employment at the two strip mills in the area (Weirton and Steubenville) had fallen from 16,000 to 12,000, but the loss at the 13 other plants had been far greater, from 16,050 to only 8,165. Three were closed and four were on part time.[33]

Sometimes the steel firms both eased their own position and managed to look vaguely magnanimous by making over the old sheet works or the site to the local community. Carnegie-Illinois Steel Corporation signed over its old works at Scottdale, Pennsylvania, to the nonprofit Scottdale Community Civic and Industrial Association for a nominal $1 as part of a program to help the mill communities. The estimated value of the land, buildings and equipment conveyed was $200,000. In 1937 it wrote off Sabraton works in Morgantown, West Virginia, and the next year donated it to the local Chamber of Commerce, which was trying to bring new jobs to the town. Weirton Steel gave the Clarksburg, West Virginia, works (the mills in which its corporate history had begun) to Clarksburg Chamber of Commerce. Carnegie-Illinois also produced maps of the layout of its derelict old mill sites, with particulars as to the services available and an invitation to any interested party to "See your broker or L. E. Jones, 1023 Carnegie Building, Pittsburgh, PA."[34] Despite such efforts, most of the old mill towns registered slow population growth or decline in the 1930s. Scottdale, New Castle and Farrell in Pennsylvania, and Dover, Cambridge, Newton Falls, Wellsville and Niles in Ohio all fell away. In New Castle, the open-hearth furnaces and tin mills were abandoned in 1937 and the Shenango works in July 1939, though it was reopened for short periods over the next three years. Together these two plants had employed over 3,000. They were considered in a case study of technological unemployment at the meeting of the CIO in 1939 and early the next year were the subject of a highly critical *Harper's* article by Harold and Stanley Ruttenburg.[35]

Behind all the company concern for modernization of production and the union battle to halt decline or at least to bring the problem to the attention of the wider community, there lay the stories of thousands of individual and family lives. Some of the best men were redeployed to strip mills. In this way Irvin works obtained a picked labor force from some 25,000 candidates, many from works closed by AS&T.[36] But such outcomes were exceptional. Much more commonly, men whose skill had allowed them to live comfortably despite their hard physical labor lost their raison d'etre or, if redeployed, had to accept deskilling. In short, behind the generalities and figures of plant closures and job losses there lay countless individual tragedies. One example was vividly described in a letter which Harold Ruttenberg received from an ex–hand mill worker in Cleveland in April 1940. Joseph Clair's letter is too long to quote here but is included in an appendix. The outbreak of war in Europe alleviated the problem, slowing closure programs and causing some mills to reopen. This proved to be no more than a delay in the relentless advance of the strip mill (see appendix).

NOTES

1. *Recent Social Trends in the United States: Report of the President's Research Committee on Social Trends*, Volume 1 (New York: McGraw-Hill, 1933), xxviii.
2. F. Purnell, "Financial problems of the steel industry," *AISI Yearbook,* 1940, 47.
3. *IA* 20 October 1938, 72, 27 October 1938, 51.
4. *IA* 30 December 1937, 55.
5. *BFSP* June 1927, 269.
6. *Steel* 16 April 1945, 82, 83.
7. *Fortune September 1931*, 48.
8. *Fortune* April 1936, 136; TNEC *Monograph 22* 1941, 111, 237, 238.
9. Stocking 1954, 115; Eppelsheimer in *JISI* 1938 II, 195.
10. *BFSP* February 1931, 261.
11. "1931 developments in sheet and tin mills," in *BFSP* January 1932, 106.
12. J. R. Miller of Cleveland quoted in *IA* 7 January 1932, 112, 113.
13. *IA* 7 January 1937, 13.
14. National Association of Flat Rolled Steel Manufacturers quoted *IA* 7 May 1931, 1545.
15. *IA* 2 January 1936, 82.
16. TNEC *Monograph 42* 1941, 62.
17. *BFSP* March 1941, 313–316.
18. Recalled by R. C. Cooper at *Hearings on Anti-Trust and Monopoly* 1957, 1137.
19. R. J. Wean quoted *TAIME* 1927–1928, Iron and Steel Section 50, 2, 6.
20. *BFSP* June 1927, 267, quoting *Mechanical Engineering.*
21. *IA* 23 October 1930, 1187, 30 October 1930, 1276, 26 February 1931, 712, 4 June 1931, 1847A; Fraser and Doriot 1932, 255–256.
22. TNEC 12 April 1940.
23. *BFSP* March 1935, 205, December 1940, 1189.
24. *BFSP* September 1939, 929; TNEC 12 April 1940, 183.
25. Glover and Cornell 1941, 409, 410.
26. Ibid., p 9.
27. Amalgamated Association of Iron, Steel and Tin Workers meeting quoted *Engineering* (UK), 30 March 1934, 329.
28. TNEC testimony 12 April 1940, 147.
29. Myron Taylor, "Ten Years of Steel," *USS* 4 April 1938, 12.
30. *H. Ruttenberg papers*. File 10. Letter of B. J. Damish, 8 December 1936.
31. *H. Ruttenberg papers*. File 10. Letters of C. Payne, 7 December 1936, and N. E. Cowen, 20 March 1939.

32. A. H. Young to W. Irvin 16 June 1936, *USS.*

33. P. Rushen to H. Ruttenberg 19 December 1938.

34. *Carnegie-Illinois* papers at USS; *BFSP* October 1938, 1021; *IA* 17 November 1938, 54.

35. Carnegie-Illinois notes; American Guides "Pennsylvania" 1940, 403,404; CIO proceedings 1939, 110–111 quoted by Carnegie-Illinois; "War and the Steel Ghost Towns," *Harper's Magazine,* January 1940.

36. Recalled in *'33 Magazine* April 1972, 54.

Chapter Eleven

World War II and the Postwar Boom

The outbreak of war in Europe in 1939 helped transform the uncertain economic recovery from the Great Depression into a period of sustained expansion of output and industrial capacity. After the United States entered the war at the end of 1941, these features became even more pronounced. As far as the strip mills were concerned, it meant a radical redeployment of plant. Light flat-rolled products had occupied prime place in the consumer-oriented growth of the interwar years; during the war years heavier products which had been in long-term relative decline once more occupied prime positions. This applied especially to steel plate for the booming shipbuilding industry. At the same time, restrictions on consumption meant that automobile and especially private car production fell to low levels. In 1941 3.78 million private cars were produced; in 1945 only 69,532. In complete contrast, whereas in 1938 the United States built only 163,000 gross registered tons of merchant shipping, 1943 output was 11.58 million tons. Sheet and strip shipments to the automobile industry in 1941 were 3.65 million net tons; by 1943 and 1944 they averaged only 0.69 million. At Carnegie-Illinois 1942 output of all categories of sheet was 1.42 million tons as compared with 2.29 million tons in 1941; for plate the respective figures were 3.94 and 2.12 million tons.[1] Under war conditions, the wide continuous strip mills were to a large extent turned over to the production of steel plate. The 96-inch strip mill at Ecorse when installed had no facilities to produce plate but even before the beginning of the war with Japan was being provided with the necessary additional equipment. By December 1941 in the immediate aftermath of Pearl Harbor, a preliminary assessment of the nation's strip mills of more than 50 inches width indicated that their capacity for sheet and strip was about 12.5 million net tons. With equipment already in place, they could produce 3.5 million tons of plate, but with additional equipment, capacity for plate might be increased to 6.5 million tons—equal to more than 85 percent as much as the nation's existing capacity of sheared and universal plate.[2] In short, strip mills spent most of World War II supplying expanded shipyards, not body shops and domestic appliance makers.

There was one other important wartime technical development relevant to the future flat-rolled products industry. By 1936, US Steel had successfully used electrolytic methods of applying tin to steel sheets in place of the traditional "hot dip" process. The new method required less tin than the old. After the wartime loss of access to Southeast Asian sources of tin, the electrolytic process came into its own. It was now reckoned to reduce the tin consumption per ton of product by not far short of two-thirds. The 1945 tonnage of electrolytic tinplate was already half that made by the hot dip process. By the late 1940s, it exceeded that number.

In the month that World War II ended in Europe, *Fortune* assessed the iron and steel industry's war record and made an effort to look ahead. The writer was disturbed by what he saw: "Steel's embarrassment of riches will be in capacity. When the war ends there will be about as much steel capacity in the United States as in all the rest of the world, and about one third more than in 1929. Probably 10 percent of the end-of-war capacity will be retired as obsolete. More will remain than the industry expects to use. The Iron and Steel Institute, adding peak pre-war demands by 19 leading customers, [only 5 of the peaks had ever come in the same year] has concluded that the best demand ever would use only two thirds of present production."[3] As things turned out, neither the forecasters at the AISI nor the writers of *Fortune* magazine need have worried. Crude steel production fell in 1946 to a level below any since 1939, but after that it resumed a general upward course. Soon it was in headlong expansion.

The nation returned to its transformation into a mass-consumption society, the process of which had been interrupted by years of depression and then by the prior demands of war. As far as sheet and strip were concerned, the major consideration was to provide the materials to satisfy the pent-up demand for new private cars. The average age of American automobiles in use in peacetime had been about 5.5 years; by spring 1945 it was almost 8 years.[4] Progress was speeded by new levels of affluence as well as by the end of wartime restrictions. 1946 output of passenger cars and trucks was 3.09 million; by 1949 it reached 6.25 million, almost a million units in excess of any previous year. The three top-selling cars of 1948 contained an average of 3,172 pounds of steel, of which more than two thirds were sheet or strip.[5] By the late 1940s, annual purchases of washing machines were about four million units, and of electric refrigerators over four million. An average refrigerator then contained almost 172 pounds of steel sheet and strip.[6] Additionally, three million vacuum cleaners, three million steel kitchen cabinets and a million electric ranges were bought annually. There was also growth in outlets for tinplate; by 1950 manufacture of beer cans alone used 400,000 tons. In that year per capita consumption of thin flat-rolled products was 414 pounds, or almost four times the level in the then most advanced of the large economies of Western Europe. In the United Kingdom, thin flat-rolled products made up 24 percent of the steel consumed; in America, the proportion was already 43 percent, according to the calculations of the Anglo-American Council on Productivity.[7]

The sustained durable consumer boom had a dramatic impact on the steel industry. When the fighting ended, it was reckoned there might be a bill of $200 million to adapt steel plants to peacetime use, and with much additional outlay to repair run-down equipment.[8] The industry then moved on from conversion and renovation to large-scale investment in new facilities. The speed of the switch and the way the momentum was kept up were impressive. Major shifts occurred in outputs, some lines of consumption important in wartime shrinking dramatically. Not surprisingly, shipbuilding was an extreme case of contraction; even in the relatively good years 1948 and 1949, steel deliveries to the shipbuilding and marine equipment sectors were well below one quarter the 1945 level. Supplies to the automotive sector were more than two and a half times as high as in the last year of war. For the "Appliances, Utensils and Cutlery" group of industries, they were 670,000 tons in 1945 and almost three times that by 1948.

Through these boom years, major steel consumers were more interested in quality of product and reliability of delivery than in price, for, though it headed a 1956 article "Steel Is the Kingpin of the Nation's Economy," the *New York Times* pointed out that the cost of steel (of all types) represented only about one tenth the price of a new car and a twentieth that of a washing machine.[9] In these circumstances, steel firms could not greatly increase their market by cutting prices. Most of the time, there was indeed no need for them to contemplate such a move. Meanwhile, consumers spread their orders between various suppliers so as to avoid becoming too dependent on any one of them. In times of labor disturbance, they would press the steel companies to come to an agreement with workers rather than have deliveries interrupted in the hope of price reductions.

A final point about flat rolled product use concerned its location. Figures of the geography of steel consumption are frustratingly incomplete; usually they do not cover all outlets, or even every state. But, both in the nation as a whole and within the manufacturing belt of the Northeast, there remained a wide imbalance between the areas of steel capacity and production, on one hand, and of consumption, on the other. For instance, at the end of the 1940s the

Pittsburgh area, which now contained 40 percent of steel capacity, consumed only 20 percent of what was produced, whereas the Detroit area with some 5 percent of national steel capacity consumed 15 percent of the total. As far as strip mill products were concerned, there were two helpful surveys by *Iron Age*—one from 1946, the other in 1960. Of 1946 receipts of flat rolled steel by metalworking plants in a group of manufacturing belt states, it was calculated that five on the Atlantic seaboard, including Pennsylvania, accounted for 38.8 percent of the total. Ohio received 18.2 percent; Michigan, Indiana and Illinois 43.0 percent.[10] Though the basis for comparison is, frustratingly, rather different, for the same groups of states 1960 estimated consumption of hot rolled sheet and strip was respectively 32.6 percent, 16.8 percent and 50.5 percent of the total. Though incomplete, these records confirm the leadership of the western part of the manufacturing belt and suggest it increased.

By early 1950 output of sheet, strip, black plate, tinplate and terneplate was at an annual rate of 27.6 million tons as compared with 15.3 million tons in the same quarter of 1940. Through the late 1940s, when there was heavy pressure on sources of supply, the industry responded by rapid extensions in its facilities. Over the three years to the beginning of 1951, capacity for hot rolled sheet and strip was increased almost 30 percent, in cold rolled sheet and strip by over 25 percent and there was a 49 percent increase for electrolytic tinplate.[11] These remained the largest group of products in the 1950s, but by then the rate of growth had slackened.

Into the next decade, there were numerous serious shortfalls in thin flat rolled product availability. There were various ways out of this difficulty. In 1947/1948 "conversion deals" were in vogue. A large buyer of sheet, unable to get all the tonnage needed, would buy ingots or even scrap and have them made into steel and rolled into sheet in any mill with idle capacity. Supplies obtained in this way might cost as much as twice the going rate.[12] Although costs of production were now reckoned to be 2.5–3.5 times those of fully continuous strip mills in the case of mechanized mills and 3.5 to as much as

Table 11.1. Deliveries of hot and cold rolled sheet and strip in relation to all shipments of steel products, 1945, 1948, 1950, 1953, and 1957

	Deliveries of sheet and strip (thousand tons)	*Share of all deliveries (%)*
1945	11,886	20.8
1948	17,192	26.1
1950	21,369	29.6
1953	23,405	29.2
1957	22,375	28.0

Based on AISI Annual Statistical Reports

5 times in hand mills, the crisis of supply gave a new lease of life to survivors of the old-style sheet industry.[13] Despite their disadvantages, acute shortages in supply meant that some old-type mills became for a time almost shuttle cocks among major consumers of sheet, often desperate for material. For example, in 1947 the Studebaker Company bought Empire Steel, whose Mansfield, Ohio, works then had a capacity of 348,000 tons of steel and 120,000 tons of sheet. The Kaiser-Fraser automobile group took over the Indianapolis sheet works of Chapman-Price and late in 1947 announced it would spend $600,000 to double its capacity. Kaiser-Fraser had a working arrangement with the integrated Portsmouth works, which was sold by Wheeling Steel when after the war it bought the Mingo Junction works from United States Steel. In Canton, the Ohio Superior Sheet Steel was taken over by Borg-Warner to supply part of the material required by its refrigerator division. GEC acquired the Niles, Ohio, works of the Mahoning Valley Steel Company, a syndicate of 25 manufacturers controlled the Apollo Steel Company, and Anderson Steel was taken by another consumer. Next year, Hudson Motor Company leased two thirds of the Shenango Tinplate Works, formerly owned by Carnegie-Illinois.[14]

The first setback to this revival of fortunes came in 1949 when, although the automobile industry remained buoyant, pressures on strip mill capacity were reduced, and they were able to satisfy more of the demand. Many old-style plants were run down or sold yet again. GEC and Empire Steel cut back sheet production, and Borg-Warner and Hudson abandoned their Canton and New Castle operations, respectively. Apollo was laid off, but when a strike was called in, other parts of the industry restarted on a temporary basis. In the course of 1949, the Chapman-Price mills were sold several times over. By the year's end, only 80–90 hand mills were left, owned by 16 or 17 companies. Some were idle, and for all of them the outlook was again bleak. At best it promised a precarious existence and a probability of sudden, maybe irreversible closure.[15]

In addition to control from outside, there were some locally significant, though again sometimes temporary, realignments among old-style sheet mills. For instance, in 1946, the Follansbee Steel Corporation of Toronto, Ohio, leased the closed capacity of the Parkersburg Steel Company to produce both motor and dynamo grade sheet. Parkersburg, with annual capacity of 36,000 tons of electrical sheet, had closed because J&L—which, when Carnegie-Illinois sold Farrell works to Sharon Steel, had been the only source of sheet bar in the region—decided to give up the trade because costs exceeded selling prices.[16] Within a few years, after the shortages had passed, Parkersburg—and Superior Sheet too—was owned by Louis Berkman, a firm which accumulated a number of redundant works. By spring 1949, another factor which had helped preserve some hand mills had changed. Continuous mills no longer rolled the thinner gauges they had produced in the earliest days. Ashland had rolled sheets down to 22 gauge, but by early postwar years the hot strip mills would not roll lighter sheets than 18 gauge. For thinner material, it was necessary to buy sheet or strip after they had been further processed in a cold rolling mill or get them from an old-style pack mill. But by the end of the 1940s, cold reduced sheet, better for many purposes than hand mill sheet, was both available and cheaper—24-inch cold reduced sheet was $4.20 per 100 pounds, whereas hot rolled, annealed and pickled sheet of the same width was $4.45. In other words, as the Pittsburgh editor of *Iron Age* put it, "For most non-integrated hand sheet mills the ten-year lease on life granted by the war is about to foreclose." Production of hot-rolled sheets on continuous mills rose 33 percent 1954–1967; for sheets rolled on all other types of mill 1967 output was only 20.5 percent that in 1954. A few old-style mills survived working on special grades and sizes. Until the early 1970s there were such operations at Zanesville, Ohio, in Newcastle, Pennsylvania, at Kokomo, Indiana, and near Chicago.[17]

There were occasional mill starts which in the long run would produce more significant results. An example was the Washington Steel Corporation, formed in spring 1945. Two years later it began to produce cold rolled stainless sheet and strip in the former works of the Washington Tinplate Company.[18] Although the evolution took many years, eventually it installed its own hot strip mill. In a few instances the transition from marginal old-style mill to strip mill operations occurred in this generally favorable time. The process usually created a fairly marginal addition to national capacity.

Although demand was booming, only three new hot strip mills were commissioned between 1945 and 1950—the finishing stands added by US Steel to the Geneva semicontinuous plate mill it had acquired from the government, in 1950 a semicontinuous mill added to the plate line at the Kaiser Fontana works and a 1948 68-inch strip mill at Sparrows Point. Apart from these, the increasing demand for sheet and strip was met by existing wide strip mills. In the short term, one way in which they coped with high demand was by cutting out much of the downtime for maintenance. Some operated for as much as 20 turns a week.[19] Many mills were re-motored, providing them with more power and greater speed, or were equipped with extra stands of rolls. Some were widened. The result of all these changes was a substantial increase in capacity. For example, by these means the rated capacity of the two Pittsburgh area mills, at J&L and US Steel, both built in the second half of the 1930s, was increased over 20 years from 1.48 to 4.36 million tons. The 1932 Cleveland strip mill J&L acquired as part of its purchase of Otis Steel was rebuilt nine years later.

The impressive commercial results obtainable from upgrading may be illustrated by the experience of Ford at River Rouge. By 1947 that company had recognized its steel operations needed modernizing to remain cost competitive. For an outlay of $54 million it managed within six years a large expansion of all the main divisions of the plant, including more than doubling its hot strip mill.[20] Generally, cost advantages were gained from "rounding out" existing plants rather than building new ones. Before the Celler Committee in 1949 Ernest Weir reckoned that the original cost of all the facilities at National Steel had been $383 million, but that to duplicate them would now cost $1.1 billion. Six years later, Arthur Homer revealed before a Senate committee that Bethlehem Steel had increased the capacity of the Lackawanna strip mill from 1.5 million tons to 2.4 million by improving the mills and heating furnaces, and that this had added only $2.9 million to the original cost of $10 million for those facilities. The present cost of a new mill of like capacity was approximately $30 million.[21] More radical improvement, even if using many existing facilities, was much more expensive. In 1926, the original hot strip mill at Butler had cost $2 million. Though larger and more powerful, a new hot strip mill which replaced it 30 years later cost $25 million.

There were various other routes to the same end of increasing supplies. Even before the end of 1945, demand for sheet was so insistent that, in a reversal of wartime practice, some continuous or semicontinuous plate mills had been pressed into use to roll sheet gauges.[22] Firms built extra stands onto heavier flat-rolled product mills so as to roll down to sheet. In some cases, either plate producers or operators of hand mills decided to install special facilities to produce continuous strip. Finally, some wholly new strip mill operations came into operation. For instance, late in 1949, a new Empire Steel Corporation was formed with headquarters in Mansfield, Ohio, which now contained its only manufacturing plant. Open-hearth capacity was increased from 350,000 tons to 500,000 tons. Two years later a 300,000-ton hot strip mill was brought into production.

Even before the war ended, a visiting senior British steel engineer recognized that his American contemporaries were moving on to a new plane. Whereas ten years before a strip mill would be designed for 900 tons a shift, the current figure was double that.[23] Of twelve new strip mills installed in this period, two (Lone Star in Texas and the Republic mill in Youngstown) were largely designed to produce material for associated tube mills, three were outgrowths from an established plate trade (Geneva, Fontana and Ivy Rock, eastern Pennsylvania). Five were ventures by relatively small companies trying to catch up with the firms which had pioneered in the trade (Mansfield, Allenport, Sharon, Newport and Portsmouth). The Detroit strip mill installed by McLouth was a brave venture by an aspiring newcomer. Only the Fairless

Table 11.2. New hot strip mills built, 1941–1957 (1957 annual capacity in million tons)

Pittsburgh Steel, Allenport, Pennsylvania	1952/53	0.72
Republic Steel, Youngstown, Ohio	1947	0.63*
Lone Star Steel, Lone Star, Texas	n.a	0.45*
Sharon Steel, Farrell, Pennsylvania	n.a	0.77
Detroit Steel, Portsmouth, Ohio	1951	0.90
Acme Steel, Newport, Kentucky	1949	0.36
McLouth Steel, Detroit, Michigian	1948	1.80
Alan Wood Steel, Ivy Rock, Pennsylvania	n.a	0.35+
Kaiser Steel, Fontana, California	1950	0.99+
US Steel, Geneva, Utah	1951	1.73+
Empire Steel, Mansfield, Ohio	1951	0.30
US Steel, Trenton, New Jersey	1953	3.12
TOTAL		12.13

* Indicates connection with tubular product
+ Indicates association with plate mills
Based on AISE 1941 and on AISI Works Directories

works of US Steel at Trenton, New Jersey, seemed to fully exemplify the new possibilities. Significantly, whereas the 1957 rated capacity of the eleven other new strip mills averaged 819,000 tons, Fairless was rated at 3,123,000 tons, making it the biggest in the country.

By the mid-1950s, the installed capacity of hot strip mills was well over three times its level at the time when the United States entered World War II. Demand had grown massively but varied a good deal from year to year. In 1955, at 38.7 million tons, or 45.8 percent of the total for all steel products, shipments of hot and cold rolled sheet and strip, uncoated and coated, black plate, tinplate and terneplate were at a peak; two years later, deliveries of the same products were down to 31.5 million tons, or 39.1 percent of all products.[24] Following long-established practice, strip mill capacity had been increased beyond even high levels of consumption; in 1955, 1956 and 1957, years of prosperity, the so-called "mill production" of what the AISI described as continuous hot sheet-strip mills averaged 39.7 million tons—roughly 75 percent of installed capacity. Inevitably there was a keen struggle for business. Corporate policies were important factors in determining success in this competition.

NOTES

1. *Steel* 16 April 1945, 82, 83; Carnegie-Illinois, *Comprehensive Plan*, 1943.
2. Russell 1941, 76; *IA* 11 December 1941, 98.
3. *Fortune* May 1945, 252.

4. *Steel* 16 April 1945, 82, 83.
5. *BFSP* September 1946, 1128, January 1949, 76.
6. *BFSP* July 1947, 812.
7. *Iron and Steel* June 1952, 5.
8. *Iron and Steel Engineer* January 1946, 54, review of 1945.
9. *New York Times International* 1 July 1956.
10. *IA* 2 January 1947, 74.
11. *IA* 10 August 1950, 87; *BFSP* October 1951, 1234.
12. Hogan *Development of American Heavy Industry in the 20th Century*. 1954, 202.
13. Armstrong 1952, 530; Fanning quoted *BFSP* August 1952, 916.
14. *Iron and Steel Engineer* January 1948, 47, January 1949, 96.
15. *Iron and Steel Engineer* January 1950, 95, 105.
16. *Iron and Steel Engineer* August 1946, 109.
17. *IA* 20 November 1947, 119, 21 April 1949, 117; *AISI Works Directory*; *IA* 31 January 1946, 21 February 1946, 91, 4 April 1946, 12, 7 November 1946, 118 and other issues.
18. *Iron and Steel Engineer* May 1947, 88.
19. Hogan, 1599.
20. *Iron and Steel Engineer* May 1953, 117.
21. Statement of Arthur Homer to Sub-committee on Anti-Trust and Monopoly, 14 June 1965.
22. *Iron and Steel Engineer* January 1946, 57.
23. Cartwright 1945, 36.
24. AISI statistics.

Chapter Twelve

Corporate Policy after 1945

In the Pittsburgh district, no new strip mills were built in the postwar years, though there were important improvements and consequent increases of capacity at the J&L Pittsburgh works and at US Steel's Irvin works. One development of the time permitted the removal of a cumbersome exchange system for materials. In 1910 J&L had made its first tinplate in the new Aliquippa works. After its first strip mill was built, the company built a cold strip mill at Aliquippa to supply the tin mills. This site obtained hot rolled coil from the Pittsburgh mill.[1] In 1956 decision was taken to build a semicontinuous mill at Aliquippa. It supplied skelp to the tube mills but also supported tinplate operations. For both finished products, all of J&L capacity was concentrated there.

Vandergrift was placed under Irvin works management in 1939 and after the war began to roll coil from Irvin. For some years, these coils were made from Vandergrift slabs, but in 1954 its steel plant was closed and rolling operations there became dependent on steel from works in the Monongahela Valley.

In this favorable period, important new mills were built in the Ohio Valley and in the interior of Ohio. Some formed parts of complicated, interlinked operations. On the Monongahela headwaters of the Ohio, the Pittsburgh Steel Company had been in existence since 1901. It specialized in wire, wire products and tubes, but during the late 1940s about one third of its steel was sold semifinished, earning a relatively low price for what was clearly a high-cost location. In 1950 the company profile was transformed by plans to enter the thin flat-rolled products field. This would involve a 48 percent increase in steel output but an 82 percent rise in finished steel tonnages. Operations began on a new 600,000-tons-per-year hot strip mill at Allenport at the start of 1953; a year later, trial operations were underway on a cold strip mill.

By 1955, 51 percent of Pittsburgh Steel shipments were in thin flat rolled products, which it had not produced before 1953. Slabs for the new mill were delivered by barge from its expanded steel plant at Monessen, five miles down and on the opposite side of the Monongahela, a link of similar type and distance, though in a different transport medium, as that between the Edgar Thomson and Irvin works.[2]

On the upper Ohio, Wheeling Steel Corporation created an even more complicated system. By 1950, it had ten works distributed along 30 miles or so of both banks of the Ohio. Their operational core was in the Steubenville and Mingo Junction works, the latter acquired in 1946. Hot and cold strip mills were at the Steubenville North Works. Beech Bottom produced electric and terne sheet; Yorkville made hot dip and electrolytic tinplate from Steubenville material.

When it acquired Mingo Junction, Wheeling Steel sold its integrated plant at Portsmouth much lower down the Ohio to a new Portsmouth Steel Corporation. By 1950, after Kaiser-Frazer had no further need of its capacity—the sheets they used now being rolled at Steubenville—the Portsmouth operations, worth $28.5 million, were purchased by Detroit Steel, whose own assets were not much more than half as large. Over the next 7 years, some $90 million was spent on developments at Portsmouth, including a 54-inch hot strip mill, commissioned in 1952. By 1957 it had 900,000 tons of hot strip and 360,000 tons of cold sheet finishing capacity, and supplied hot rolled coil to cold reduction mills in Detroit. A somewhat similar form of backward integration was followed when the small melting shop and sheet plant at Newport, in the Kentucky section of greater Cincinnati, was bought as a source of material by Acme Steel of Riverdale, Chicago. From 1949, it was equipped with a small strip mill. Portsmouth and Newport were to be of marginal importance in the strip mill industry, in size, product and by virtue of a less-than-favorable location for production and marketing. But there was at this time one more important instance of entry, and the beginnings of a second. Both were in favorable locations and soon made good their initial lack of a full cycle of iron/steel/rolling mill operations.

Acme Steel acquired Newport in September 1956, but in the next year it began a process of expansion which would render this link superfluous. It decided to spend $23 million to install at Riverdale oxygen-blown converters backed up by hot blast cupolas. Projected capital outlay for 450,000 tons of steel capacity was only $51/ton. In the mid-1960s, Acme merged with Interlake Iron, which owned a blast furnace plant in the southern outskirts of greater Chicago, and the new Interlake Steel Corporation became a fully integrated strip mill operator, though always on a small scale.[3]

Until this period, the Detroit area contained only two large and integrated iron, steel and hot strip mill operators—Ford at River Rouge and the Great Lakes Division of National Steel at Ecorse. A third was now added. McLouth Steel had been incorporated in 1934. By 1949, it was bringing into production a new steel plant at Trenton along the Detroit River some 10 miles south of the older and very much bigger complexes of the Rouge and Ecorse. Trenton's annual capacity of 430,000 tons raw steel depended on electric furnaces bought from the YS&T Indiana Harbor works. Through the early fifties Trenton was relatively small, a not fully integrated producer of flat-rolled products, but by now it possessed an electric steel plant with a capacity of a little under one million tons and a 600,000 ton reversing sheet mill. In Detroit it operated a reversing hot strip mill of 108,000 tons and cold strip mills. Further growth highlighted the leading part the needs and plans of the automobile industry could play in the industry. The Ford Motor Company had doubled its strip mill capacity, and General Motors, having no steel operations of its own, decided it might be faced with supply shortages. As in the early days of mass auto production, it resolved to contribute to expansion by steel companies in return for a commitment to provide some of its requirements. Accordingly, it put up $40 million to help finance steel extensions by Republic and $28 million for J&L to increase open-hearth capacity and flat-rolled output in Cleveland. The agreement for the latter provided for repayment by 1966 in the form of monthly installments. GM agreed to buy almost all the increased output of the works.[4] It also bought $15 million of Algoma Steel debentures, and had conversations with McLouth. Perhaps as important as McLouth's close proximity to the southern Michigan operations of GM was the fact that as early as 1952 this small company had examined the new LD oxygen steelmaking process in its Austrian homeland. It had no commitment in open-hearth capacity which would raise the costs of an early entry to the new process. The outcome was that GM bought $25 million of McLouth preferred stock and gave it a contract to supply 5 percent of GM domestic requirements. McLouth used the enhanced standing GM's commitment had given it to raise another $78 million, and by 1953/1954 had decided to go for full integration—an ore dock, iron making and new steel and rolling capacity. By 1957, the Trenton works had a blast furnace, a three converter, 540,000 ton oxygen steel shop—the first BOF plant in the United States—and a 1.8-million-ton hot strip mill. 1954 sales were $59 million; nine years later they reached $234 million, about half of which were in business with General Motors.[5]

A relatively small but increasing part of the industry was concerned with higher-grade strip mill products. For such companies, advantages or disadvantages of location were of little significance, for effectively they were operating in a different market from the bulk producers of carbon steel flat-rolled products. Allegheny-Ludlum Steel had been formed in 1938 by the merger of the western Pennsylvania Allegheny Steel and Ludlum Steel of New York State. From 1932, Allegheny had operated a wide strip mill at Brackenridge, the former site of West Pennsylvania Steel, 20 miles up the Allegheny from Pittsburgh. When World War II began, this was the smallest of the nation's 26 strip mills. In 1950, the company decided to build a new mill. A few years later, its chairman summed up the special commercial circumstances under which his company operated. Its products, including flat-rolled steel, were generally priced at about $600/ton when those from carbon steel makers were then less than $200. Indeed, he pointed out, "In the specialty steel industry, we measure our shipments in pounds rather than in tons. . . . In our flat rolled plants the average order is about 16 tons [and in these plants] I am sure we have fifty different alloys in process at any one time."[6] Crucible Steel was another quality entrant into sheet and strip. In the late 1940s, it built a 66-inch reversing hot mill and cold strip mills at its Midland works on the upper Ohio. This was mainly concerned with stainless flat-rolled products. In the early 1960s, it installed a continuous strip mill. As well as the quality firms, some bulk steelmakers also had an interest in these special fields. The Canton and Massillon works of Republic's "Central Alloy District" had a small hot and cold rolled sheet and strip capacity, and when in the late 1950s Republic decided to build a new strip mill at Warren it was designed not only for carbon but also for stainless and silicon sheets. US Steel had produced electrical-grade sheets at Vandergrift from 1902. It now centered more of this and of its stainless sheet business there, increasingly drawing hot-rolled coil from Irvin works to be cold rolled on a Sendzimir mill.[7] Butler, the pioneer hot strip mill, was not an integrated operation and was gradually developed as Armco's special flat-rolled products plant. It had first rolled stainless strip from Middletown ingots in 1930 and by 1939 began to produce aluminized material. Nine years later, electric furnaces were installed so that it could make its own stainless and alloy ingots. By the mid-fifties, when a decision was made to replace the thirty-year-old strip mill with a larger, more powerful unit, most of its sheet and strip was in stainless, electrical and coated carbon grades. A few years later it was decided to almost double the Butler steel capacity.[8]

In 1956 and 1957, J&L was another major company to gain a footing in sheet and strip of special grades, buying both the Cold Metal Products Company, which had cold mills in Youngstown and Indianapolis, and Rotary Electric Steel of Detroit. A year later, it installed a new stainless strip and sheet plant at Louisville, Ohio. The annual capacity of 36,000 tons indicated that such specialized operations, though interesting and profitable, were a sideline to the competitive array of major strip mill operators.

Table 12.1. Capacities of prewar strip mills rebuilt, 1955–1967 (million tons)

	1941	*1955/1956*	*1967*
Bethlehem Steel			
Lackawanna	0.81	2.4	2.5
Sparrows Point	0.67	1.7	2.7
Ford, Dearborn	0.39	1.3	2.0
Inland Steel, E. Chicago	0.67	1.3	1.7
Jones & Laughlin, Pittsburgh	0.81	1.8	1.5
National Steel			
Ecorse	0.94	2.4	2.4
Weirton	0.72	1.8	3.1
Republic Steel			
Cleveland	0.90	1.9	2.2
United States Steel			
Irvin	0.67	2.5	3.1
Fairfield	0.50	1.3	2.8
Youngstown	0.47	1.1	1.3
Youngstown S&T			
Youngstown	0.84	1.2	2.5
East Chicago	0.67	1.5	2.0
TOTAL	9.06	22.2	29.8

Based on Ess 1941, 1970; Cordero 1955

An important route of entry to the sheet/strip business in the early postwar years was from plate. The US Steel works at Homestead and South Chicago had both been provided with continuous light plate mills in the mid-1930s, but from World War II onward, they operated wholly in the plate business, no longer appearing in listings of the nation's strip mills. Now, however, this route was followed by some companies in peripheral regions of the national economy, areas in which, though total consumption might be large, it was comparatively widely scattered as compared with the mass outlets in the western parts of the manufacturing belt. For instance, the 48-inch strip mill installed by US Steel at Fairfield in the late 1930s had shared roughing stands with mills rolling universal plates, and when in 1957 Republic belatedly decided to break into the regional market for thin flat-rolled steels with mills superior to the sheet mills at its Gadsden works, the 800,000-ton semi-continuous strip mill it installed there used "breakdowns" from the plate mill. A similar pattern was followed in the West. At both Geneva and Fontana, plate and strip mill operations were closely associated, strip gauges being rolled on additional stands downstream from the plate mill. It was a relatively cheap way in which to equip a mill to cope with demands from two very different directions, neither of which, alone, could command markets adequate to support the whole plant. Both works were helped by other factors. Kaiser Steel was allowed to use wartime profits, which otherwise might have been heavily taxed, to pay off part of the RFC loans it had been given, and US Steel benefited from the low capital cost at which it acquired the works it had originally built and operated for the government to supply wartime demands, particularly for ship plate. In 1946 it bought Geneva for 25 cents on the dollar of its original cost.

On the Atlantic seaboard, the situation was more favorable than in the West or in the South for strip mill expansion. Even so, some schemes proved abortive. Perhaps the most visionary, but also vague, was the idea and campaign for a fully integrated works and the production of thin flat-rolled products in New England.[9] A relatively small development was undertaken by the old family firm of Alan Wood. Long a factor in the plate business, it now installed additional stands to a plate mill so that for the first time it could produce strip/sheet gauges, although only in narrow widths. A study for a completely separate hot strip mill there was commissioned, but there was no positive outcome.

The situation of the new US Steel Fairless works on the Delaware was different. This was a major greenfield plant built at the inflated costs of postwar construction—but in the midst of the Korean War boom. In a few years it would prove to have been additionally burdened by the fact that all its newly built raw steel capacity came from open-hearth furnaces, for in a period much shorter than that of any reasonable amortization, this proved an outmoded technology. When to these disadvantages is added that of a lower-than-projected growth of the eastern seaboard metalworking economy, it may be appreciated that Fairless would eventually disappoint the high hopes expressed at its inception.

Despite the varied entry routes into strip mill operations, between 1940/1941 and 1956/1957 far and away the major source of expansion of capacity came from upgrading mills which had been installed during the preceding period of almost identical length since the continuous process had been introduced. The 22 existing works increased capacity in this section of their businesses by a tonnage two and a half times as great as that of the ten strip mills added in new plants. "Rounding out" continued to be much cheaper than completely new construction. For instance, having only entered the strip mill trades in 1953/54, within seven years Pittsburgh Steel decided to extend its Allenport mill. At a projected cost of only $4.9 million, it expected to increase hot mill capacity there by 20 percent, from 750 to 900 thousand tons, by adding two additional stands to the existing four, enabling use of heavier slabs and producing heavier and wider coils.[10] The focusing of increased capacity in existing units is particularly striking in contrast with the pattern of the next round of expansion which would involve major and completely new units, though again in works already having strip mills.

This period also provided evidence of the strengthening attractions of a location on the shores of either Lake Erie or Lake Michigan, but these largely involved new plants or projects. In contrast, for major firms with works both in the interior and on the lakeshore, there was as yet no very definite trend in favor of the latter. In fact, between 1941 and 1957 the combined capacity of strip mills at five interior works of US Steel, Republic, J&L, Youngstown Sheet and Tube and National Steel was expanded by those companies marginally more than that of their lake works. (The mills involved were Irvin, Warren, J&L Pittsburgh, Campbell [Youngstown] and Weirton in the interior, and in lake shore areas Gary, the Cleveland works of Republic and J&L, Indiana Harbor and Ecorse.) The experience of this period also proved that locational advantages were permissive, not automatic. In other words, in some cases the intrinsic benefits of a favorable location might lull those who controlled operations into complacency whereas in apparently ill-favored situations, management could be spurred to find ways making them acceptably profitable. In short, as always, the final arbiter of relative success or failure was quality in management. The early postwar experiences of two leaders in the industry illustrated both aspects of this situation.

National Steel Corporation's Weirton plant, like the interior mills of a number of companies, concentrated on tinplate, but its superbly located Great Lakes Ecorse works had been designed from the start to deliver bulk supplies of strip and sheet to nearby automobile press shops. By 1954, 97.8 percent of its hot rolled steel capacity was of either sheet or strip. George Fink had been the Great Lakes Steel president since the plant was commissioned 24 years before. He was now sixty-eight. The year 1954 was one of sharply decreased output of both passenger and commercial vehicles, and in these difficulties the controllers of Great Lakes made a critical mistake in overestimating demand from their main customers. For the whole steel industry in that recession year, the value of products shipped and of services was 19.6 percent lower than in 1953, and net income declined 13.3 percent: at National Steel, the respective falls were 24 percent and 38 percent. 1954 total revenue at National Steel was 76.3 percent of the 1953 figure whereas at US Steel, Armco and Inland it was 84.2 percent, 90.8 percent and 92.8 percent respectively. The next annual report of National Steel was quite explicit as to where the blame lay; their relative failure had been a consequence of the company's long years of all-too-easy profit making: "Under the new conditions created by this reduced demand for steel, we discovered that a number of practices had developed in the operation of this subsidiary which were entirely out of harmony with the principles on which National Steel conducts its business.

Table 12.2. Capacities of major company mills in the Northeast interior and on the shores of the Great Lakes, 1940–1973 (thousand net tons)

	1940	*1957*	*1964*	*1973*
Five "interior" works (YS&T, Youngstown; USS, Irvin; J&L, Pittsburgh; National, Weirton; Republic, Warren)	3,472	9,298	11,613	11,470
Five lakeshore works (YS&T, Indiana Harbor; USS, Gary; J&L Cleveland; National, Ecorse; Republic, Cleveland)	4,125	10,835	15,345	

Based on AISE 1941; AISI 1957, 1964; Cordero 1974

There was an excessive accumulation of inventory, insufficient maintenance, an undue amount of labor disturbance and other deficiencies in operation. This had not come to notice prior to that time because operating conditions resulting from the lengthy boom in steel demand not only led to the gradual development of these faulty practices but also obscured them." In response, a "thorough change in management personnel, policies and practices and a complete reorganization was put into effect." Fink was retired from the presidency; a year later he left the board.[11]

The experience of Inland Steel in the Chicago area was generally a happier one, though here again it was the complications of the policy choices of individuals and of boards that provided the essential medium for interpreting possibilities and tackling problems. In 1947, Inland spent $16.1 million mainly for modernization and a 300,000 ton extension in its facilities for thin flat-rolled products, but it was generally cautious in its early postwar expansion program. Whereas its near neighbor, YS&T, having decided it would move from the situation in which roughly two thirds of its capacity was in Youngstown to a roughly 50:50 Youngstown/Chicago split, increased the capacity of its Indiana Harbor mill by almost 127 percent over the few years after 1947, Inland, whose only plant was in that district, expanded by less than 71 percent in the same years. Between 1951 and 1954, rapid amortization was made available by government to encourage steel expansion, but Inland spent less than any other major company. It was not until early 1956, by which time costs were again rising, that it launched a $300 million expansion program. It considered increasing steel capacity by installing the new basic oxygen process, but, lacking sufficient blast furnace capacity to provide the necessary molten iron, let slip what would have been a pioneering move for one of the majors and instead added three 315-ton open-hearth furnaces. Yet, although in these respects it may seem to have been unimaginative, in other ways Inland's performance was good. In 1957, when seven other majors, all with larger raw steel capacity, averaged an 86 percent operating rate, Inland's figure was 100 percent. By spring the following year, when, at 53 percent, the industry was operating at a lower level than at any time since 1938, Inland was utilizing 74 percent of its capacity. Its advantages of closeness to markets were important, but by no means a sufficient explanation of this success, for in the first quarter of 1958 the Chicago-area plants of US Steel, Republic and YS&T were all running at 10 percentage points below the Inland rate. An attractive consideration with local customers may have been that it was the only wholly local major supplier, but consumers are commercially oriented and not liable to be overly swayed by sentiment. More important, Inland seems to have nursed its market. As a major buyer of sheet steel was reported as saying in the late 1950s, "At Inland you deal with individuals not with a branch office of a faraway outfit. When we have a problem we can go down and talk with the boss."[12]

With the benefit of hindsight it seems that, in terms of an efficient distribution of its capacity, J&L misdirected its expenditure on its major plants and above all on its strip mill operations in the early postwar years. It spent heavily on modernizing Pittsburgh operations, even though, as in the massive developments twenty years before, this involved large outlays to clear ground. In the first two years of the fifties, the company planned to spend $228 million as a result of which steel capacity at Pittsburgh would increase by 1.2 million and at its Cleveland works by only 360,000 tons. However, by the late 1950s, the emphasis was moving away from Pittsburgh. The company pioneered large-scale basic oxygen steelmaking at Aliquippa in 1957 and from 1961 operated the process at Cleveland too. Ominously, at Pittsburgh J&L held on to the open-hearth process as this passed rapidly from dominance to obsolescence, though it must in mitigation be recorded that a large part of its capital outlay a few years before had been in a new 2.1-million-ton melting shop. Pittsburgh contained 60 percent of the J&L hot strip mill capacity in 1957 but less than one third seven years later.

During the quarter century to the mid-1960s US Steel and Bethlehem Steel built three new centers of strip mill operations, two in the eastern and western peripheries of the national economy and the other at its core. In 1937/38 the Chicago district contained 1.4 million of US Steel's 2.7 million tons of annual strip mill capacity; by 1957 it had only 3.2 million out of 13.1 million tons. Providing an interesting sidelight on the implications of its location, in the 1950s as new strip mill capacity came into operation, the strip mill at the Irvin works became more focused on orders of the hard-to-roll type.[13] Bethlehem Steel controlled 15.3 percent of the nation's hot rolled products capacity in 1957, and 14.1 percent of the rated capacity of its strip mills, but had no capacity in the Chicago area, though it was by then making purposeful moves to make good its lack. In this period Republic, third in size among the majors, showed how varied and flexible development might be in a complex concern. In the late 1940s operations at its 42-inch Warren mill were severely overtaxed in supplying coil for tinplating. The company reacted by grafting strip mill production on an 84-inch Youngstown plate mill, and by adding three 4-high 48-inch finishing stands and coilers, and so forth. This took over the business of rolling up to 30,000 tons a month of skelp for the company's three Youngstown-area electric weld tube mills and 20,000 tons of tinplate band for its Niles works. Within a few years, the Youngstown operation was classified as a strip mill, but by the sixties, when shortages of material were no longer pressing, it went out of production.[14]

Armco continued to do extremely well notwithstanding some problems. By the early postwar years, it operated seven steelmaking plants. Three, controlled by its Sheffield Division west of the Mississippi, played no part in the thin flat-rolled products industry. Baltimore was a small, specialty works. Its three wide continuous sheet/strip mills, were, in increasing size, at Butler, Ashland and Middletown, which also accounted for 63 percent of its 1957 raw steel capacity. Each was in the interior and only two were integrated. It was reckoned that the small average size of all the Armco plants—in 1957 850,000 tons—meant that its overall costs in making steel were higher by about $10/ton than those of Bethlehem Steel. Yet it more than compensated for its liabilities with the high overall value of its products. As a result, in the difficult trading conditions of the first half of 1959, though the smallest of the "big eight" in raw steel capacity, Armco ranked fourth in sales and profits. In net income per ton of steel shipped it was first. Its largest plant, Middletown, obtained its hot metal by rail from the Hamilton blast furnaces 10 miles away. The Middletown strip mill, originally installed in 1928 and rebuilt in 1937, was again partially rebuilt in 1958. By the early 1960s, its annual capacity was 2.25 million tons. Ashland was modernized, a process that involved rebuilding 80 percent of the operation. An important part of this process followed a 1950 decision to spend $15 million converting its continuous mill which rolled individual sheets into a continuous wide strip mill turning out coil.[15] Butler presented other problems, being small, unintegrated and landlocked. The solution was to specialize in much higher-grade products for which such disabilities were of little concern.

In this period, some individuals who had played a pioneering role in the early strip mill industry passed on. John Tytus, honored as vice-president technical development at Armco, died in summer 1944. His colleague, Charles R. Hook, gave up the Armco presidency in 1948, but remained chairman of the board for another 11 years. Tom Girdler retired as chief executive of Republic Steel in 1955, aged 78. In spring 1957, Ernest T. Weir gave up the chairmanship of National Steel. He was in his 82nd year and died within two months.

NOTES

1. *BFSP* August 1949, 947.
2. "The new Pittsburgh Steel Company" n.d; *BFSP* August 1953, 928; A. Adams, "Pittsburgh Steel Company: A Progress Report" 1956, 7; *BFSP* July 1946, August 1954.
3. *BFSP* September 1957, 1045.
4. *IA* 14 December 1950, 122.
5. AISI *works directories*; *Fortune* January 1965, 235; McLouth annual reports.
6. *BFSP* June 1950, 699; E. J. Hanley, "The Specialty Steel Industry," Notre Dame University, 20 October 1966.
7. *Iron and Steel Engineer* June 1959; *BFSP* December 1953, 1467, August 1959, 888, 889.
8. *Iron and Steel Engineer* August 1961, A2; *Steel* July 1964, 35.
9. See Broude 1963, 110–121.
10. Pittsburgh Steel, annual report 1961.
11. National Steel, 1954 annual report, 8, 9; *Fortune* October 1957, 156; *AISI annual reports*; see also Kaplan et al., 167, 205.
12. *BFSP* April 1948. Supplement, 84; "Inland does it again" in *Fortune* July 1958, 95, 96, 100; *Iron and Steel Engineer* June 1960, 98.
13. *'33 Magazine* April 1972, 54.
14. *IA* 19 June 1947, 109; *AISI works directories.*
15. *Fortune* November 1959, 132; *BFSP* August 1962, 757, October 1950, 1214, January 1964.

Chapter Thirteen

The "Second-Generation" Mills

In 1941, when the Association of Iron and Steel Engineers published *The Modern Strip Mill*, there were 28 hot strip mills—defined as continuous operations rolling strip of 36 inches or more in width. Their annual capacity was 17.6 million tons. (Two, which accounted for 1.6 million tons of the total—Homestead and South Chicago—were confined to the lighter gauges of plate, and subsequently were not generally classified as strip mills.) By the early 1960s, estimated rated capacity of strip mills had increased over 250 percent to 63.6 million tons of sheet, strip, coil and black plate for tinning and a small proportion of light plate.[1] Expansion on this scale was partly part due to construction of new mills, the total number having risen by this time to 42, but it was also in considerable part the result of rebuilding of older mills. Both new installations and rebuilt units were provided with bigger slabs, mill drives which had been re-motored to increase their power and rolling speeds and other though relatively minor improvements. From the end of the fifties there began another forward surge, involving construction of another group of mills. Incorporating outstanding, distinctive properties they were called "second-generation" strip mills. Twelve were installed in the 14 years to 1974, but in many respects their prototype had been completed in the mid-1950s, as a major rebuild rather than a wholly new plant.

In 1927, Weirton Steel Company commissioned a mill to roll strip 48 inches wide and with a rated monthly capacity of 30,000 tons. In 1947, the finishing train of this mill was rebuilt and its strip width increased to 66 inches. As a result of this and other improvements, the rolling speed was increased to 1,600 feet per minute and capacity to 171,000 tons a month. In 1955, acting for Weirton Steel, the United Engineering and Foundry Company carried out a more fundamental rebuilding of the same mill. The width of the finishing train was reduced to 54 inches. An additional drive motor was fitted to each stand, three slab reheating furnaces, the largest in the world, were installed, and the roughing train was completely rebuilt. Delivery speed in the mill increased, this time by 50 percent to 2,400 feet per minute. Rated monthly capacity went up to 225,000 tons, and the material was delivered in much larger coils than in the old mill—up to five times in length—for further processing into sheet and tinplate. This effectively "new" mill not only turned out bigger tonnages but also brought advances in product quality and important operating economies. A leading expert concluded, "The entire project is an example of good engineering and ingenuity."[2] A few years later, he recognized it had been the harbinger of another revolution in strip mills.

As had occurred on many occasions, the improvements now made were a response to pressure from consumers for a better product, involving larger coils rolled to closer tolerances. The incentive for each company to invest in the new mills was to win operating economies, but also to respond to the challenge from rival firms which had already done so. Naturally, many of these plants would themselves later be upgraded, their capacity in many cases increasing as a result. There was a danger that their high capital cost might cripple them financially, a potential threat that would become a harsh reality if consumption did not continue to grow rapidly or, more specifically, not sufficiently quickly to allow them to benefit from high utilization rates for their extremely costly units of plant.

By the late 1950s, Republic Steel was operating a one-million-ton 42-inch mill at Warren, Ohio, which, like that at Weirton, had been installed in 1927, but afterward was modernized and widened to its present size.[3] Republic replaced this mill in 1961 with a 58-inch mill of 1.5 million tons capacity—twelve years later it had been expanded to 2.1 million tons. For mills making wider strip, and keenly contending for the bulk orders of the automobile industry, a breakthrough came in 1961 with the starting of an 80-inch strip mill at National Steel's Ecorse works. Over the next 13 years, ten more strip mills were brought into production. All but the last were of 80-inch or 84-inch width. New mills of 80 or more inches in width had a 1970 combined capacity of 30.4 million tons. Their construction set off a transformation in thin flat-rolled products with important ramifications throughout the rest of the 20th century.

Whereas over the years the capacity of earlier strip mills had been added to and their products improved gradually, spending on more powerful drives and increasingly sophisticated controls meant that the second-generation mills were superbly equipped from the start. More specifically they were provided with much more electrical power than former mills, which greatly increased mill speeds and therefore output. They were computer operated and equipped with sophisticated gauge controls. Heavier slabs were used, and they had bigger slab heating furnaces, equipped for the first time with extractors designed to reduce wear and tear in the process of removing the slab from the furnace. Their cranes had much greater lifting power than those in earlier mills. Technically the second-generation mills were a triumph, but their unquestioned achievements were won at high expense.

There were additional costs to take into account, for the mills required backup from other items of new plant. These included cold reduction mills on the downstream side, and supplying them involved not only outlay on slabbing mills but also more steel and in some cases additional blast furnace capacity. These in turn required more ancillary plant and calls on resources at each stage of the process of manufacture. As a higher standard of product was at once achieved—at any rate after solving any running-in problems—they helped make older strip mills into second-rank suppliers. By implication, if consumption fell away—at least for the more demanding outlets—the earlier mills would become marginal suppliers. One or two examples highlight these problems.[4]

Responding to current demands from its customers for "wider, heavier, weld-free coils of the highest quality," in 1961, Inland Steel announced plans for a new 80-inch hot strip mill to be built over the next three years. Stockholders were assured that "no present mill will surpass it in size, power or speed." When brought into production in spring 1965, the new 102,250-hp Indiana Harbor strip mill was well equipped to meet exacting market standards. The first mechanical extractors removed slabs for the hot strip mill from the reheating furnaces and delivered them to the roughing stands without the usual "gouging and other damage to the bottom surface." Moni-

Table 13.1. The "second-generation" strip mills

Year	*Company*	*Location*	*Width*	*1973 Cap . (th.ts)*
1961	Republic	Warren	58″	2,100
1961	National	Ecorse	80″	3,600
1964	J&L	Cleveland	80″	3,060
1965	Inland	Indiana Harbor	80″	3,000
1965	Wheeling	Steubenville	80″	2,400
1966	Bethlehem	Burns Harbor	80″	3,500
1967	YS&T	East Chicago	84″	3,000
1967	Granite City	St Louis	80″	3,000?
1967	US Steel	Gary	84″	4,420
1968	Armco	Middletown	86″	4,000
1970	Republic	Cleveland	84″	3,042
1974	Ford	River Rouge	68″	2,900
	TOTAL			38,022

Based on Ess 1970; *Iron and Steel Works of the World* 1974

tored and controlled by a computer system, the new mill could produce coils weighing up to 83,600 pounds—as compared with 30,000 pounds in the new Weirton mill only ten years before—and these coils were of more consistent gauge, width, metallurgical characteristics and better surface quality than those from Inland's older mills. Operating costs were reduced by a then-unique roll-changing system, making it possible to replace worn rolls in all six stands more rapidly than previously in only one stand. These features indicated the revolutionary nature of the new mill. Yet from the start, as Inland made clear, this major unit of plant was only "the first step in a vast new long range expansion program."[5] The new mill would be sited on land reclaimed from Lake Michigan and near to cold reduction mills built between 1955 and 1958. Wider implications in relation to capital outlay and for the competitive situation became clearer over the next few years. First it was decided to install two 230-ton BOF converters—at an anticipated cost of $200 million—which would mean closing the No. 1 open-hearth melting shop, whose origins dated from the 1901 first developments on the site. Experience soon justified this extra outlay, for the new plant could make steel at a saving of $5/ton over the capacity it replaced.[6] At the commissioning of the hot mill, it was pointed out that, when required, more reheating furnaces and material handling equipment would permit an increase in the effective capacity of the mill, and space was provided in the layout for additional slab heating furnaces capable of doubling the initial mill output. Investments already made had increased the Inland capacity substantially.

For example, the BOF shop not only replaced open-hearth furnaces but also brought a net increase of one million tons in raw steel capacity.[7] The company already had two hot strip mills, of 76-inch and 44-inch width, with a combined capacity of 3.27 million tons. The new mill added 2.4 million tons, or over 70 percent, to Inland's hot rolled coil and sheet capacity. By 1970, rated capacity at the new mill was raised to 2.8 million tons, and by 1973 to 3.0 million. Before the end of 1970, an additional 80-inch cold reduction strip mill was commissioned. Such major and interlinked extensions increased the status of Inland Steel, already a more than usually efficient and successful operator. But what would happen if—or, more appropriately, as—other companies made similar commitments in new mills—particularly if they served the same market area, and even more if demand ceased to rise, or its rate of increase faltered?

Those who installed the first new mills were effectively throwing down a challenge to the rest of the industry. Some companies were much less well placed than Inland to make a successful response. Their existing share of the market, plant location or finances could provide barriers to success. Wheeling Steel was disadvantaged in all three respects. When the American Iron and

Steel Institute issued its 1957 Works Directory, Inland hot strip mill capacity was 3.3 million tons, that of Wheeling 1.3 million. In 1960, the latter had been eleventh among the majors in both steel capacity and output, and its operations were divided among a number of places on the upper Ohio. Net income per ton of steel shipped was $6.52 at Wheeling and $11.92 at Inland; from every dollar of sales the former had a net income of 3.63 cents, the latter 6.3 cents.[8] Yet over 10 years Wheeling Steel spent $163 million on plant additions and improvements. In November 1962, it announced a three-year expansion and improvement program. Of this, $30 million was for "normal" replacements, additions and improvements, but, following a five-year-long intensive study by its own staff as well as by consulting engineers, Wheeling committed itself to $145 million for new facilities and equipment, including an oxygen steel plant to supplement the output from its open-hearth furnaces, and an 80-inch mill in place of its existing 66-inch strip mill. This new mill would initially have a monthly capacity of 200,000 tons but might be expanded to 300,000 tons. That this venture was a remarkable act of faith was shown by the fact that in no year until then had Wheeling Steel shipped more than 1.54 million tons of products—of all types—and the average for 1960, 1961 and 1962 had been 1.25 million tons. To finance the project, the company arranged to sell $105–$120 million in First Mortgage Bonds over a three-year period and agreed with five major banks for a revolving short-term credit of $20 million.[9] The new Steubenville mill was brought into production in 1965, but the effort expended in building it seemed to exhaust the company. There were serious internal dissensions and losses in both 1965 and 1966, years in which Armco for instance recorded record production and profits. By 1966, Wheeling was unable to obtain loans for further modernization. Next spring, the president and vice president of Pittsburgh Steel took over at Wheeling. A year later the company was merged with Pittsburgh Steel, whose assets were only about half as big as its own.[10]

Three other major companies which built second-generation mills also lost their independent existence. Granite City Steel in 1960 ranked 15th in steel capacity. In 1967 it commissioned a 2.4-million-ton strip mill, but shipments of steel products failed to grow, as a result of which it had "neither the market nor the finishing facilities to support a full load on its mill." This was fatal for a highly capitalized operation. In the three years preceding that in which its new mill was commissioned, Granite City had an average net income of 6.13 cents for every dollar of net sales; over the three following years net income was only 0.9 cents per dollar.[11] In August 1971, it was absorbed by National Steel which in 1970 had made 12.1 cents on each dollar of net sales.

Having completed a new cold reduction mill at Indiana Harbor in 1965, Youngstown Sheet and Tube decided to install an 84-inch strip mill there.

It planned to finance some 80 percent of the program, including ancillary plant, from internal resources. Eighteen months later, the decision was made to build a BOF plant. Following commissioning in 1968, difficulties were experienced in operating the new mill and in the next year company shipments were 5 percent lower than those of the year in which it had decided to build. Early that year, Youngstown revealed it would be acquired by the Lykes Corporation, whose interests were in banking, insurance, shipping and electronics.

A final loss of independence among the builders of second-generation mills involved an older, bigger concern, Jones and Laughlin, which in 1960 ranked fourth in national steel capacity. In 1964, it brought on stream a 2.4-million-ton strip mill at Cleveland works. The cause of corporate failure lay elsewhere, though in a related development and in what at first had seemed an imaginative stroke in development of its strip finishing capacity. In June 1968, after a distinguished history of more than one hundred years, J&L was acquired by the conglomerate LTV.

As well as contributing to commercial failures and helping to cause important mergers, the second-generation mill program highlighted a further shift in locational preferences. There was only one brand-new, greenfield site, and the plants receiving the new investments made by the 11 companies and involving over 32 million tons of strip mill capacity were wholly within the northeastern manufacturing belt, but they were very strongly oriented toward the shores of the Great Lakes. In contrast to the previous 20 years, no new mill was installed on the Atlantic coast, or in the Southeast, Southwest or Far West. Five companies had existing mills both on the Great Lakes and on the upper waters of the Ohio or in the landlocked area between that river and the Lakes. One built a new-style mill in the interior—Republic at Warren—but this was designed to roll special products, not to compete in bulk carbon steel strip. With this exception, the five companies located their new capacity on the lake shores, totaling 15.9 million tons. There was also a strong westward trend. Of the 32.4 million tons of new strip mill capacity, 8.7 million tons was on Lake Erie or near Detroit, 13.4 million tons at the southern end of Lake Michigan. In 1957, 15.3 percent of the wide strip mill capacity controlled by US Steel and by Bethlehem Steel was in the Chicago district, a proportion more or less identical with that in the industry as a whole; ten years later the share of this district in the total strip mill capacity of the two industry leaders was 24.7 percent. For the industry it was 19.0 percent.

Cleveland investments by two companies with important hot mill capacity in Pittsburgh and Youngstown reaffirmed the locational attractions of the lower Cuyahoga Valley, notwithstanding its badly crowded site conditions. Ford, the last of the companies to make a move to new-style mills—in 1974, and therefore not included in the above figures—inevitably chose Detroit, but if it had wished and judged it desirable, National could have built instead at Weirton. The Chicago situation was even clearer. Between 1965 and 1968, four new-generation mills with 13.4 million tons of capacity were located in the short stretch of Lake Michigan shoreline from East Chicago to Burns Harbor. Inland, like Ford, was a one-plant company with no realistic option of investing elsewhere. But US Steel chose to locate its major investment at Gary rather than at Irvin, and Youngstown Sheet and Tube opted for East Chicago, not Youngstown. Most dramatic of all, the magnetism of Chicago at last brought Bethlehem Steel into the Midwest, after a corporate history of over 60 years, during half of which it had contemplated such a move.

In the early 1930s and again in the mid-1950s, Bethlehem had attempted to merge with Youngstown Sheet and Tube, but on both occasions it was defeated by court action. It had made clear that, though the link it was seeking would give valuable access to the Ohio Valley, and from there to the whole Mississippi basin and the Southwest, its main aim was to acquire a footing in the Chicago area through the YS&T East Chicago works. After its second defeat, Bethlehem prepared to make its own way into Chicago. Availability of a suitable lakeside site in the sand barrens of Porter County, Indiana, well to the east of Gary, had been drawn to its attention in 1930. Shortly after World War II—and almost a decade before the second merger attempt failed—Bethlehem began an exhaustive search for a Midwest site. (This started as US Steel was surveying the possibilities of an eastern seaboard location from which it could invade a regional territory until then dominated by Bethlehem.) In 1956, 1,200 acres on the shore of Lake Michigan was purchased, the holding was later increased, and by 1963 Bethlehem had embarked on an initial building program costed at $250 million at what now became known as Burns Harbor. Plans were announced for two 200-ton oxygen converters, a sheared plate mill, a cold reduction mill and tin mill. Later that year the company ordered a hot strip mill.[12] In the early stages of operations, the works depended in part on slabs railed from Lackawanna, but it was clear that, though temporarily an additional outlet for the latter, Burns Harbor constituted a longer-term threat to the older plant's very existence. When less-favorable market conditions arrived, the Lackawanna strip mill, although on the lakeside, would like those in Youngstown or Pittsburgh be an encumbrance to a company operating a second-generation mill so much nearer the prime markets of Detroit and Chicago. By 1973, strip mill capacity at Burns Harbor was already almost 40 percent greater than at Lackawanna. Expansion continued in the Chicago area plant, and the Buffalo operations were gradually discontinued.

There were at this time two other important indicators of the way things were moving in terms of locational response to the markets, though in the end neither proved anywhere nearly as decisive as Burns Harbor. In October

1929, the newly formed National Steel Corporation had bought a site at Portage on Lake Michigan, west from the future Burns Harbor site. For a time, it planned a major plant to complement its Great Lakes Steel unit at Ecorse. The Depression frustrated this scheme, but 30 years after its acquisition National decided to build cold reduction mills at Portage, to operate under the name Midwest Steel. It hoped these mills would reduce the instability in its existing operations caused by wide swings in the highly specialized southern Michigan market by providing access to the more diverse Chicago outlets. Initially Portage would use hot rolled coil from Ecorse, but eventually it was intended that its own facilities would be filled out by backward integration, making it a complete iron and steelworks. By 1964, A. J. Berdis, president of Midwest, reckoned more than half the tonnage it shipped was business which National Steel could not have obtained without it. Late in 1966, work began on a deepwater dock, given the name Indiana Lakeport, making it possible in the first instance to deliver coil from Detroit by water, but, as the annual report put it, "Ultimately, the port will make possible Midwest's integration as a steel making plant."[13] This process was never completed.

Another important indicator of the increasing attraction of production in the Midwest came at the same time from J&L. It had been periodically reported as interested in Chicago-area development since the 1920s but there had been no positive outcome. After emphasizing plant modernization throughout the company under the long chairmanship of Ben Morell, and a cut back on capital outlay during the short period of control by Avery Adams, from 1963 J&L embarked on another phase of expansion. It greatly expanded the hot strip mill at its Cleveland works, and in 1962 decided to spend $60 million to install one of the new-generation mills there—even though in order to provide it with a 10-acre site it was necessary to realign the Cuyahoga and remove 600,000 tons of earth. This new Cleveland project marked a major shift for J&L lakeward. Two years later, with construction of this mill already far advanced, it examined possible sites in several Midwestern states for a new cold rolled and galvanized sheet plant. Like National, it chose to depart from the practices until now common in almost all the big companies of placing the hot and cold mills in the same works, and instead, in 1965 announced it had chosen a riverside tract at Hennepin on the Illinois River in the interior of Illinois. As with National's Portage operation, this would in the first instance finish hot rolled coils from existing plants, in this case from Cleveland, Pittsburgh and Aliquippa, for the last two of which it would provide some mitigation of the disadvantages of interior locations well away from major outlets for finished flat-rolled products. As with National at Portage, there were longer-term plans. The huge size of the Hennepin site—6,000 acres, about twice the area of that at Burns Harbor on which Bethlehem was building a fully integrated complex—was the best indicator of the J&L scale of thinking. Something was made of the possibility that part of this tract might be occupied by customers, but the main interest centered on prospects for backward integration. The company tried to prevent speculation of an early start, but in fall 1966 the J&L president admitted, "Someday, we expect that Hennepin will be our fourth integrated steelworks." Changing circumstances defeated this aim. It became known that the company decided on the project even though its own studies showed that most steel buyers who planned to relocate intended to put their new plants in Kentucky or Indiana, and not as far west as central Illinois. Moreover, outlay of $172 million on this mill meant that modernization of high-cost operations elsewhere in the company were stalled for want of cash.[14]

In the late twenties and during the thirties each major company had sooner or later felt compelled to build one or more strip mills to maintain its standing in the industry. Now, thirty years later, they were striving to the same end with second-generation mills. In each case there was no practical alternative; they had to build or would lose their place in the trade. But, as always, there was a danger that supply would get seriously out of step with demand. A number of factors brought this from the sphere of hypothetical threat to that of painful reality. The main considerations, more fully examined elsewhere, included a less steady growth of the automotive industry, considerable economy in use of flat-rolled steel per vehicle, serious shrinkage in at least some of the important outlets for tinplate and a considerable increase in imports of sheet and strip. To these changes was added the background consideration of much less certain national economic growth. If demand fell, the industry might drift into a crisis in which second-generation strip mills would survive but many older mills would go to the wall. Even before this, threats of regional overcapacity were foreseen, especially in the Chicago area, the focus of so much of the expansion. In 1963, J. L. Block of Inland Steel noted that two expected newcomers—presumably Bethlehem and National—would increase hot rolled steel capacity in the area by 10–12 percent. He believed that he could reassure stockholders, for their own analysis had indicated that over the next decade consumption of steel in Midwest states served by Chicago mills would increase by about 25 percent, with an even greater rise in light flat-rolled products. Five years later, a conservative estimate put expansion in the hot strip mill capacity in the Midwest represented by the four new Chicago area mills and that at Granite City at 15 million tons, with a potential further increase to 20 million tons.[15] Twenty years further on, by which time both regional and national overcapacity had become a serious problem, each of the new mills not only survived but also had expanded. But some of the older strip mills, at the same locations, or even more those in less favorable situations or which had not been modernized, would be closed and abandoned. As with the first strip mills, the second generation swept away its predecessors.

NOTES

1. AISI "The Modern Strip Mill," preface; *IA* March 1962, 54.
2. National Steel 1955 annual report pp. 2, 3, 7; T. J. Ess, "Weirton Steel Company," *Iron and Steel Engineer* November 1958, 12, 14.
3. *Iron and Steel Engineer* June 1959.
4. Hogan, 1601–1609.
5. Inland Steel report dated 21 February 1962, 8.
6. J. L. Block to *New York Society of Security Analysts,* 1967, 8.
7. *Steel* 30 August 1965, 62, 64; Inland Steel, "The 80 inch hot strip mill," in Inland annual report 1965, 12, 13.
8. *IA* 6 April 1961, Steel Industry Financial Analysis.
9. Wheeling Steel 1962 annual report.
10. E. Nolin and B. E. Seeley in Seeley, 1994.
11. Granite City 1970 annual report; National Steel 1971 report, 5.
12. *BFSP* January 1963, 46; June 1963, 480, November 1965, 1051; *IA* 4 February.
13. *Fortune* February 1964, 214, "National: A pacesetter in steel": National Steel annual report 1966.
14. *BFSP* July 1962, 666; W. J. Stephens, "On making waves," speaking in La Salle, Illinois October 1966; *Wall Street Journal* 5 January 1970, 1, 9; *Business Week* 6 Jun 1970, 110–112.
15. Inland Steel annual report 1963, 2; 4 March 1968, 7.

Chapter Fourteen

Years of Difficulty

The 1970s and 1980s

During the 1970s, the steel industry passed from high and generally rising levels of activity to uncertainty and then into the early stages of a major crisis. Except for tin mill products, the thin flat-rolled section became an increasingly important part of the whole industry. The next decade was one in which the steel companies had to undertake the most draconian rationalization of the twentieth century. By then, even their strip mill sector was suffering severely.

The seventies began well enough. Rated strip mill capacity reached a peak in the early part of the decade, but the increase was not evenly spread. Between 1967 and 1973/4, there was a 6.4 million tons increase in national capacity but declines of about one million tons in the recorded capacity of the areas within 50 miles of Pittsburgh and in the East. There was a smaller fall in the West. Throughout this period, industry spokesmen were, outwardly at least, bullish about prospects. In spring 1967, speaking in San Francisco, the president of US Steel, Leslie Worthington, forecast a 40 percent increase in the steel market over the next 10 years. Five years later his company projected an increase in cold rolled sheet demand from 16 million tons in 1970 to 25 million by 1980.[1] Time proved these optimistic projections wide of the mark.

At the end of 1970, raw steel capacity was reckoned—by the *Wall Street Journal*—to be 197 million tons. The industry itself opted for a much lower figure. Output that year was 131.4 million tons. Two years later, production was almost 151 million tons, and the industry reckoned it would need to install 25 million tons of new capacity by 1980. The chairman of Bethlehem Steel announced his company was prepared to spend over $2 billion for 6.6 million tons of capacity over 12 years, and early the following year, the Inland Steel chairman reckoned industry-wide expenditure for seven years would need to be $25 billion. In the event average annual output of raw steel from 1975 through 1979 was almost 14 percent less than in 1973 and 1974. Thoughts of expansion turned into programs for reduction. By one assessment, between 1977 and 1986 raw steel capacity fell from 160 million to 128 million tons. Cold rolled sheet deliveries were 14.2 million tons in 1970 and 17.3 tons in 1979, but generally the strip mill sector was involved in the process of pruning capacity.

Through the sixties and into the early seventies, some 36 million tons of new strip mill capacity had been built. In a period of approximately equal length, the 15 years from the completion of the last of the second-generation mills, that at River Rouge, through to 1989, no new, bulk product, hot strip mill was added. The only additional units were two 56-inch-wide mills installed by Washington Steel Corporation in western Pennsylvania to produce hot rolled stainless coil for its cold reduction mills; their combined capacity was a little over 400,000 tons. Two major projects were announced. In 1974, National Steel was reported to be planning to complete by 1980 a two- to three-million-ton-per-year integrated works near Gary—the long projected rounding out of its Portage finishing operations with iron- and steelmaking capacity and a hot strip mill. It was not built.[2] For a time, the United States Steel Corporation was involved with an even bigger scheme. In 1973, Edgar Speer became the USS chairman. Under his direction, the Corporation actively planned a major new integrated works on a greenfield, lakeside site at Conneaut, Ohio. Its centerpiece was to be a three-million-ton hot strip mill, which it was planned would be increased by 50 percent in a phase 2 of development in the late 80s. Through the difficult years of the late 1970s, Speer persisted with his vision of a vast new Lake Erie mill. Ill health compelled him to retire in spring 1979, and his successor as chairman quietly dropped the whole scheme. Circumstances were also changing rapidly for existing plants and companies, even for some well placed in terms of location. As late as 1978, *Iron Age* wrote, "In terms of new technology, no steel company in the world is more advanced than McLouth Steel." Three years later, McLouth filed for Chapter 11 protection.[3]

Though the high cost of new capacity in a very uncertain market ruled out new mills, the upgrading or modernization of existing mills could be an important aid to their survival. Of the 61.5 million tons of capacity in 1989 in mills able to roll strip of 56-inch width or more, units accounting for 19.2 million tons to some degree or other underwent modernization at this time. In some cases, the improvements were fairly small, as, for instance, in 1977 at Armco's Butler works, where a fourth slab furnace was involved. Specialization there on stainless and electrical grade sheets enabled the plant to survive through succeeding years of difficulty when much bigger, integrated but ordinary grade producers failed. For a number of works, strip mill capacity by the end of the 1980s was considerably smaller than in 1973/4. Usually, this was because of the retirement of mills of earlier date, leaving second-generation mills in the same works to meet the full or a much larger share of demand. For this reason, the combined capacity of the strip mills at Inland Steel, Great Lakes Steel, Ford, Gary, and Steubenville fell in 15 years from 28.9 to 19.8 million tons. There were smaller declines at Warren and Middletown, little change in the two Cleveland mills and at East Chicago, and a modest expansion at Burns Harbor. National Steel made a larger extension of strip mill capacity at its Granite City operation. The most conspicuous contraction at a plant with no second-generation mill was at Sparrows Point. In 1973, the two strip mills there had a combined annual capacity of 5.2 million tons; by 1989 the older, narrower mill had been closed and the 68-inch mill, originally installed in 1947 and modernized in 1964, remained at its 1973/1974 rated capacity of 2.7 million tons.

Between 1973/1974 and 1989, eight mills (and all or most of the integrated iron and steel operations connected with them) closed. This removed some 11 million tons of strip mill capacity. One of the two largest of the casualties was Lackawanna, but its cold reduction mill and associated galvanizing operations survived. The other seven mills lay either on the Ohio and Monongahela or in landlocked sites between them and the shores of the Great Lakes. Although during these fifteen years national rated strip mill capacity fell by 25.5 percent, in the Detroit area there was no change overall, and capacity in the Chicago area went up by 2.3 percent. In marked contrast, there was a fall of 27.0 percent on the Lake Erie shore—above all due to the closure of the 2.5-million-ton Lackawanna mill—and 42.8 percent shrinkage in the greater Pittsburgh district. For the Valleys, the decline was just short of two thirds, Youngstown losing both of its main mills, the Campbell works of Youngstown Sheet and Tube and US Steel's Ohio Steel Works and McDonald mills. On the middle Ohio, capacity fell by almost a third. Armco's Ashland works more or less held its own, but in 1980 Interlake Steel closed its Wilder (Cincinnati) works. Opened again the next year under worker ownership, its hot strip mill now supplied material for the tubular goods trade. Eventually, not long after it finally closed the strip mill operation in 2001, the company signed a contract with Nucor to supply it with hot rolled coil.[4] Decline at the bigger, fully integrated Portsmouth works was even more striking. In 1981, Empire-Detroit Steel shut it down and eventually the remains of steel operations there were cleared to make room for a Walmart store. Change on the upper Ohio was less dramatic, but the strip mill Pittsburgh Steel had installed at Allenport was abandoned in favor of the larger Steubenville operation after the merger which produced Wheeling-Pittsburgh.

As the massive decline in capacity indicated, there were important changes in the Pittsburgh area. Vandergrift never had a strip mill, and since 1954 had depended on Irvin works for its material. Despite falling volume, it was moderately profitable until the 1986/87 steel strike, in which it lost so many of its customers for silicon sheet that afterward volumes were low and costs so high that it operated at a loss. By early 1988, this works, which had once employed 4,000, had jobs for only about 100. In 1988 it was sold by USX to Allegheny-Ludlum, whose operations were nearby in Brackenridge and West Leechburg.[5]

The Irvin works itself faced some difficulties in this period. In the early 1970s, at a time of high activity, slab supplies from Edgar Thomson had to be supplemented from Gary and Fairless works, but shortfalls in supplies in this period sometimes held new cold reduction operations at Irvin to 12 turns of the potential 20 in a week. A decade later, full operation of the Irvin hot mill was said to require 2.7 million tons of slabs a year, but Edgar Thomson could only deliver 1.9 million tons. For a time, Duquesne and Homestead made up the shortfall, but soon they were recognized as strong candidates for closure, and the company had to face up to installation of a high-cost slab casting plant at ET. Not until 1989 was a decision taken to go ahead with this. The capacity provided was double that of a unit considered in 1982, but the projected cost had increased from $42 million to $250 million.[6]

The never-ending struggle to survive in the difficult conditions of the late seventies and the eighties was illustrated even in Chicago operations. The 80-inch strip mill at Gary, built in 1936, was modernized in 1948. In 1967, US Steel installed one of the new-style strip mills, an 84-inch unit, at Gary. Availability of this unit meant that the 80-inch mill—now the "old" mill—operated only to meet peak demand. After 1974 it was unprofitable, and although $7.6 million was spent on upgrading it in 1976, in July 1979 it was decided to suspend operations, and six months later to dismantle it. The overall cost of closure was projected to be $14.5 million, but it was anticipated that on the profit-and-loss account the annual saving would be $15.5 million. Pressure to improve Gary operations continued. By 1985, the 84-inch second-generation mill had already produced 54.8 million tons of hot rolled band—an annual average of 3.04 million tons as compared with its estimated initial capacity of 3.5 million tons. But by early 1985, USS felt it necessary to conduct a comprehensive study to order to determine what was needed to make this mill a "world class" unit in terms of productivity, quality of product and reliability. Remarkably, it was found that for an annual outlay of $686,000, yearly operating and maintenance costs could be cut by $2,956,000.[7]

In the East, Southeast and West, the strip mill industry more or less held its own, but there was no sign now of the progress these peripheral areas had registered in early postwar years. As shown above, strip mill capacity at Sparrows Point was almost halved, the Fairless mill in 1989 was rated at

71 percent the capacity of 1973/4 and the small Alan Wood works had given up its tentative attempt to get into the medium-width thin flat rolled product market. In the South, the Fairfield works survived and grew slightly, but in 1986 LTV, which now owned Republic Steel, sold the Gadsden mill to its employees. The two bright hopes of the West noticeably faded. Fontana, shorn of supporting iron and steel operations, turned to use of purchased slabs, underwent various modernizations, and its rated capacity was slightly increased. Geneva, still an integrated operation, was no longer in the forefront of Corporation thinking. Capacity shrank; in January 1987 it was resolved to "idle indefinitely" operations, and by 1989 it had been sold to Geneva Steel.

Company alignments changed substantially in this period. In 1983/4 US Steel offered $575 million for National Steel but dropped its bid when the Department of Justice indicated that it regarded such a merger as breaching antitrust policies. Later that year Nippon Kokkan bought a half share in National for $273 million. In 1983, having reduced the proportion of its steel needs which it provided for itself, the Ford Motor Company closed the River Rouge operations, which afterward were conducted by Rouge Steel. A more complex process was involved in the evolution of LTV Steel. In 1968 this conglomerate acquired an 80 percent share in J&L.

In terms of company development policy, there was, whenever a choice was possible, striking unanimity in choosing to make extensions or to hold the line on the Lakes rather than in the interior, and to prefer the western rather than the eastern part of the Great Lakes basin. Bethlehem Steel's major complex at Lackawanna at the eastern end of Lake Erie was a delayed but predictable casualty of its decision to force its way into the Chicago area. For access to the heart of the Midwest industrial area, Burns Harbor was much better located than its earlier foothold on the Great Lakes, and, as marketing conditions worsened, it was logical that the more peripheral plant would be run down to ensure as full a load as possible for the new, highly efficient but also high-capital-cost, core-area mill. J&L followed the same logic. Having built a second-generation mill in the Cuyahoga Valley, and committed itself to a major cold rolling complex along the Illinois River, it closed its Pittsburgh and Aliquippa mills. Youngstown Sheet and Tube, which had built a new mill at East Chicago, closed its Youngstown mill—which resulted in a $138 million charge against income—and US Steel, focusing on Gary, ended operations in Youngstown, began a rundown at Fairless and barely held the line at the Irvin works. National Steel reaffirmed its position in Detroit but, after the increasing use of aluminum reduced outlets for tinplate in cans, and as competition became keener, announced in 1982 that it would no longer make major investments in Weirton operations. Two years later, it sold this works to employees.

Table 14.1. Shipments by the steel industry, 1970, 1974, and 1979 (million net tons) and share of thin flat-rolled products

	1970	1974	1979
All products	90.8	109.5	100.3
Thin flat-rolled as % of total			
Hot and cold rolled sheet and strip	32.0	33.8	35.4
Tinplate and other tin mill products	8.0	6.9	6.3
Galvanized or other coatings	6.7	7.3	8.0

Source: AISI Annual Statistical Reports

The many adjustments, reductions and closures of this period meant that another social crisis began. It was serious but not as dramatic as the earlier crisis caused by strip mill displacement of jobs in the old hand mill industry. When J&L reopened its Pittsburgh works for a time early in 1972, it planned to cut 2,200 jobs, enabling it to make 10 percent more steel with 30 percent fewer workers.[8] Sometime later a new president and chief executive, Thomas C. Graham, aimed to further improve J&L performance by a sharp reduction in employment. Moving to US Steel in 1983, he pursued a similar policy there. For many years there had been a saying in the Valleys district: "When Detroit is down, Youngstown is flat."[9] Now the closure of both the US Steel and the Youngstown Sheet and Tube operations caused a large and permanent loss of steel jobs. When US Steel decided to close both its integrated Ohio works in Youngstown and the linked McDonald bar and strip mills, the annual cost saving was projected to be $7.0 million. The combined loss of employment was anticipated as 3,913.[10]

In summary, in what was in effect the national periphery for this industry—the Atlantic seaboard, the Southeast and the West—it had been noteworthy that there had been no investment in the second generation of strip mills. The unquestioned heartland of the strip mills remained in the western part of the manufacturing belt. Here too there was a distinct pattern of changes. Seven locations having only first-generation mills ceased to operate strip mills in this period. At older-style mill locations which survived, there was an overall decline in capacity, which was especially marked in those plants which had also installed second-generation strip mills in the sixties or early seventies. Even at well-located second-generation mills, the increase in capacity was slight.

NOTES

1. *IA* 11 May 1967, 11; *Iron and Steel Engineer* January 1972, 80–82.
2. *Wall Street Journal* 7 December 1973; *Mining Annual Review* 1975, 74.

3. Quoted "A retrospect of 20th century steel," *New Steel* November 1999.
4. *Cincinnati Commercial* 29 July 2003.
5. USS *CPC* 16 November 1987; *Pittsburgh Post-Gazette* 3 June 1988.
6. *Pittsburgh Post-Gazette* 24 November 1971; USS *CPC* 6 December 1982.
7. USS *CPC* 7 January 1980, September 1985.
8. *Metal Bulletin* 4 January 1972, 35.
9. *New York Times* 27 April 1958.
10. USS *CPC* 7 January 1980.

Chapter Fifteen

EAF Steel and Compact Strip Mills

In striking contrast with earlier years, through the last quarter of the twentieth century consumption of strip mill products increased only slowly. This was especially the case with the automobile industry, as always the most vital factor in the market. Private car production in 1991 and 1992 averaged a mere 0.73 percent more than in the three years 1967–1969, though commercial vehicle output went up 143.2 percent. The new emphasis on smaller cars meant that the importance of the industry as an outlet for home strip mills fell away. However, over the terrible years for steel in the 1980s and into the 1990s, the strip mill industry changed dramatically. By the mid-1990s, shipments of hot and cold rolled strip and of tinplate had declined sharply, but whereas cold rolled sheet output also fell, shipments of hot rolled sheet increased. Most impressive of all, there had been a large rise in shipments of galvanized sheet and strip, made both by hot dip and electrolytic processes, as automobile makers used them as anti-rust materials. Imports of sheet and strip products increased.

Many strip mills were reconstructed during the 1980s. Of 17 installations of more than 50-inch width by 1991, eight had been modernized since 1980. The process was carried further through the nineties with major investments in plants of varying age, size and national importance, including Burns Harbor, Sparrows Point, Granite City, Indiana Harbor, the Rouge, Riverdale and Warren. Now, however, commercial justification for these improvements was threatened by increased competition between established mills, larger imports, and particularly by the early stages of a wholly new homegrown strip mill industry. This last threat came from the "mini-mills."

Mini-mills, steelmaking operations based on electric arc furnaces and in which primary rolling mills are replaced by the continuous casting of steel, had been a growing sector of the steel industry for many decades. In 1960,

10–12 of these mills shared about 2 percent of national steel capacity. By 1984, they had increased to about 50 in number and held between 15 and 20 percent of the market. Yet despite their increased general significance, at that time they seemed to represent no threat to the strip mill sector, for as an account of the time put it, "The products made in the rolling section of a mini-mill are simple and limited in variety. They include wire rods, reinforcing rods . . . and bar products in a variety of forms." They might also roll light beams, channels and angles. But by this time further changes could already be anticipated: "It also seems likely that many mini-mills will broaden their range of products and the quality of the steels they make."[1] Even now they provided no direct competition for strip mills at integrated works, which could call on a steady supply of large slabs of assured quality, but indirectly mini-mills had already affected the sheet and strip business. Their advance in the fields of bars, rod, light structural steel, etc., had made major companies increasingly dependent on their hot strip and cold reduction mills. At the end of the eighties, there came the first direct assault on this core area of the integrated companies. Long before the nineties ended, mini-mills were recognized as constituting a serious threat to integrated strip mill operators.

There are three main material bases for cost saving in mini-mills as compared with integrated iron, steel and rolling mill plants. Firstly, they short-circuit the traditional production cycle by cutting out blast furnaces and all the associated manufacturing and mineral operations on which large-scale iron making depends, simply replacing them with large electric furnaces working on scrap. Secondly, they have been using continuous casting from a time at which the bigger producers still depended on primary rolling mills to work their ingots down into slabs, and this helped them in the form of both reduced capital outlay and lower process costs. Finally, they finish their steel in thoroughly up-to-date rolling mills. In the 1960s, most mini-mills were small—from 60 to 200 thousand tons a year. Later they increased in size.

As the listing of their advantages suggests, financially the biggest source of economy in mini-mills as compared with integrated operations is in capital outlay, but operating expenses are also lower. This is due only in part to the reduced cost of inputs; there are additional, organizational factors. The new-style technology has been accompanied by a sharp reduction in manning levels, more flexible management and less of the traditional divide between blue- and white-collar workers. They generally employ non-unionized labor. Sometimes, largely because of their common choice of location in a rural area, wage rates may be relatively low, but often employees are paid as well as those in the integrated industry and the lower unit labor costs result more from absence of job demarcation and a general flexibility of outlook and practice than from savings in the wage packet. For these various reasons mini-mill productivity is high. By the early 1990s, labor productivity was often of the order of 200–300 percent above that in many integrated works, and this was reckoned to give them a cost advantage of as much as $100/ton.

Dependence on scrap has been recognized as a potential Achilles' heel for mini-mills but even here wider changes within the industry have helped them. The general advance in continuous casting throughout the steel industry meant that much less process scrap was produced than when it depended on primary rolling mills, but this was counterbalanced by the collapse of the open-hearth steel process, which removed an important rival buyer for scrap. A general rule is that the trend of scrap prices provides a check to mini-mills when steel markets are buoyant but gives them a boost in times of lesser demand. In contrast, the cost curve for integrated works is flatter. For a time in the 1990s when scrap prices at length seemed likely to rise to inconveniently high levels, mini-mill companies campaigned, though without much success, to restrict scrap exports. At the same time, recognizing the danger signs, they began to seek out alternative methods of supplying their iron requirements. These alternatives included investment in the production of directly reduced iron (DRI). Sometimes, by venturing into pioneering technologies, they suffered setbacks, as with direct steelmaking processes based on iron carbide. Inevitably such investments pushed up capital outlay per ton of product, reducing their advantages though providing a safeguard against inflation of scrap prices.

Boosted by a range of technical and commercial advantages, mini-mills extended their operations into products previously beyond their range. By the end of the 1990s, no major integrated plant retained important heavy structural capacity; it had all been lost to these smaller rivals. They entered rail production. Most threatening of all, at last they moved into and then rapidly increased their standing in the production of sheet and strip. They did so by adding to their range of continuous cast products that of thin slabs suitable for direct rolling down in strip mill stands.

Although J&L engineers are said to have worked for a time on the development of thin-slab casting and seem to have made important progress, they eventually gave up work in this field.[2] The breakthrough to commercial success came from a German plant equipment maker. After considering entering the production of thin flat-rolled products for some years, at the end of 1986 Ken Iverson, head of the mini-mill firm Nucor, recognized that this "SMS process," using a continuously cast "thin slab," and proved in Germany in a pilot plant by 1985, provided the key to a successful venture into strip sheet.[3] Pioneering this on a commercial scale, Nucor began to build a new works on a greenfield site at Crawfordsville, Indiana, in fall 1987. This began operations in late summer 1989. Within less than a year, it was operating profitably. Using a 100 percent scrap burden, Crawfordsville was equipped with two electric arc furnaces and a continuous casting unit which produced "slabs" 2 inches in thickness. These passed directly to a newly built 4-stand, 4-high strip mill. In the earliest days, although expensive reheating and some of the intermediate steps in which quality might suffer were cut out, the new plant could only make lower grades of strip, and consequently representatives of the integrated industry endeavored to appear unruffled. Yet by 1991, although the general tone of a review written by a former research director of USS remained lukewarm, it revealed a vague anticipation of an increasing threat from this infant industry: "For many years, the sheet business, which at its high price end is very technologically intensive, was inaccessible to the minis. However, a courageous gamble by Nucor that they could cast slabs a few centimeters thick by the SMS process and thereby avoid conventional hot-rolling paid off after considerable start-up problems. From this start, they were able to make sheet which is adequate for some purposes. It appears that further development is likely to be able to capture increasing fractions of the sheet market in the same way that much of the bar market was swallowed."[4]

Nucor, and soon afterward other mini-mill companies, were confident they could follow this course of advance—gradually moving from lower to higher grades, thereby capturing increasing shares of production. Some of their innovations were, to a layman at least, complex or even incomprehensible. A key step in their progress involved the introduction of methods of "calming" the meniscus in the mold so as to give improved control over slab quality, reducing both metallic inclusions and the surface blemishes which had until then had excluded the use of mini-mill sheet or strip for the most exacting purposes.[5] Their entry into the strip mill business began to turn into a surge. By late 1993, three companies—Nucor, a newly formed mini steel firm, Steel Dynamics, and a Canadian group, Dofasco—were reported as already operating or building of the order of 6 million tons of mini-mill sheet capacity. As one engineer pointed out, this more or less equaled the capacity of the Gary strip mill—it amounted to a major new competitor in a market well supplied by existing operators. By 1995, $3.5 billion had been invested in flat-rolled product mini-mills in the United States and Canada. They turned out only 3 million tons of hot rolled bands and other thin flat-rolled products, but then reckoned that within another five years they might ship 14 million tons.[6]

Experience proved that savings on the capital and production accounts are immense when high-grade thin flat-rolled products are made in mini-mills. By early 1996, the cost-reduction potential of the new technologies was put at 35–40 percent. In 2000, one leader of the mini-mill sector claimed that whereas integrated mills had an average operating profit of the order of $20/ton, the profit of his company, Steel Dynamics, was $60/ton. At the same conference it was reckoned that most existing mills needed to cut their costs by \$50–\$70/ton to maintain their competiveness, and that for most of them savings on this scale were impossible.[7] Given such circumstances it became common for mini-mill men to denigrate the old-style industry—Iverson once spoke of the merger of Republic and J&L into LTV as "like tying together two anchors to see if they would float," and a few years later an executive of IPSCO described integrated mills as being in "full paralysis."[8] Yet it also began to be recognized that in this struggle established producers possessed important compensating advantages. They looked ahead to a time when changes in the trend of scrap prices might prove unfavorable to their new competitors if there was a general switch to electric furnace steel. Ahead of this hoped-for boost, they continued to spend heavily to upgrade their own operations even though profits remained small. The productivity gap between integrated and mini-mill operations was narrowed, so that the best of the former could match the profitability of the latter. Some older companies reequipped themselves so as to gain some of the advantages possessed by their nimbler rivals, and in this way divisions between the sectors of the industry became blurred. For instance, in 1995 Armco started up a thin slab caster at its Mansfield works. Late in 1996, Acme Steel at Riverdale, Illinois, planned a \$370 million project for a new continuous thin slab caster and hot strip mill of 900,000-ton capacity. Designed to draw on the company's blast furnace and BOF converter facilities, it was expected to cut manufacturing costs by as much as \$70/ton.[9]

Notwithstanding their brave efforts, integrated strip mill operators generally were worsted in the fight against their leaner opponents. In the late nineties it was reckoned that in making hot rolled merchant grade band mini-mills operated at about 0.75 man-hours per ton as compared with 1.5mh/t at the best integrated works—though the latter were even now better equipped to turn out value-added products, such as cold reduced band. In 1994, US Steel had a net income of 3.31 cents per dollar of sales; Nucor made 7.63 cents.[10] But some "old-style" producers did manage to match the returns of the mini-mills, as was shown by the fact that over a number of years the operating profit per ton at AK Steel exceeded even that at Nucor. In short, the alleged cost advantages of the mini-mill sector were by no means as clear cut or conclusive as was often claimed. Yet a variety of circumstances suggested that eventually keener competition would lead to a severe reduction in the number of integrated strip mill operations.

Firstly, expectations that high electricity and scrap prices would frustrate further progress in the mini-mill industry proved ill-founded. By the end of the century, electricity supplies to EAF furnaces cost no more than about 3 cents per kilowatt hour.[11] For a long time, projections of the inhibiting effect of higher scrap prices seemed wide of the mark. After rising in 1995 to more than the $130/ton threshold, heavy melting scrap fell away sharply over the next four years. By summer 2000, it ranged as low as in the years of deep depression in the decade to 1987.[12] Meanwhile the development of DRI and other sources of iron, including the recognition that "low" blast furnaces—as opposed to what one critic emotively described as the "mastodons" of older, bigger furnaces—might be excellent pig iron producers, requiring lower grade and therefore cheaper minerals, further eased fears of crippling failings in metal supply. The mini-mills received another boost when use of large amounts of oxygen was seen as a way to reduce tap-to-tap times in their furnaces, increasing the productivity of the capital tied up in plant, by perhaps doubling its effective capacity. Electrically made steel consequently further improved its standing. As late as 1992, a year after the last open-hearth furnace was abandoned, steel made by BOFs accounted for 62.0 percent of all tonnage as against 38.0 percent for electric furnaces. Seven years later the respective percentages were 53.8 percent and 46.2 percent.[13]

As the cost situation of mini-mills remained generally favorable in the late 1990s, national steel output was no longer increasing and prices fell. This was marked in flat-rolled products. From summer 1997 to late 1998, the world spot export price for hot-rolled band fell from $350 to $185 a metric ton. By spring 1999, some bigger producers had secured large four-year contracts to supply General Motors, but at prices giving them continuing low incomes.[14] In the hope of improving their standing in the market and eventually reducing costs, integrated producers continued to make heavy investments in their hot strip mills and in new or extensions to cold rolling and coatings plants. Mini-mills were making their way into the higher-grade flat-rolled products, but large outlays were also made at, for instance, the USS Gary and Irvin works, at the Ecorse, Granite City and Portage operations of National, by Inland Steel, now included in the ISPAT group, and at Armco and Wheeling-Pittsburgh. By summer 2000 the chairman of Bethlehem Steel claimed his company had reduced costs "over the past few years" (not only in the strip mill sector) by as much as $300 million by eliminating underperforming assets and restructuring pension costs.[15] In 1998 and 1999 alone four other integrated producers, none of the first rank—Weirton, Gulf States, Geneva and WCI—spent $186 million on modernizing operations.[16] Yet yield prices continued to be highly volatile. Most ominous of all for longer-term prospects, the upgrading of plant usually caused an increase in capacity and thereby increased the keenness of the rivalry for future business.

Given competition from expanding capacity at mini-mills, from overseas and from each other, the big producers were hard hit in times of low demand and collapsed prices. As seen, by summer 2000, the head of Steel Dynamics claimed that across all products his own company had an operating profit of $60/ton compared with an average of $20 for integrated companies. Now enjoying 20 percent of the steel industry's volume, he believed they could increase this to one third or even more. By 2000, Nucor was involved in capacity extensions which would make it bigger than the United States Steel Corporation.

Summer 2000 speakers at an analysis of the industry and its possible strategies for survival recognized that overcapacity would probably mean more combinations and almost certainly some failures. A representative of Rouge Steel projected that in perhaps eight years consolidations might mean there would be no more than four or five strong companies. The economic consultant, Donald Barnette, reckoned that to maintain their competitive position, big integrated companies would have to cut costs by $50–$75/ton, would not meet this target, and that the result of their failure would be restructuring, bankruptcies and lower capacity. Another contributor recognized that capital spending had meant that "the United States is swimming in excess capacity."

The likely outcome of all this seemed to be that over a number of years the number of strip mills at integrated works would fall and capacity would be concentrated either at the few which survived or in mini-mills. Already there were signs of disturbances in the trade along these lines, but there was also a continuing reluctance of corporate players to withdraw from the struggle. Their tenacity caused yet more competitive pressures and more changes. In December 1999, Weirton Steel was sold. In the mid-1990s, Caparo Steel purchased the former Sharon Steel operations at Farrell for $26 million. It resumed production on the 600,000-ton hot strip mill. Later another group was formed to buy and continue this plant and modernize its 60-inch hot strip mill and the cold rolling mill. By summer 2000, though in Chapter 11 bankruptcy, Acme Metals was still operating its compact strip and thin slab mini-mill, and planning to spend $26 million on necessary replacement works. Geneva Steel, though bankrupt, made large investments. Gulf States Steel, whose spending program had been largely intended for environmental improvements, began to close its operations in August. Generally, however, the stakes continued to rise. In 1999 and 2000, unintegrated mills—in all lines of their business, not just strip mill products—spent or planned to spend $3.16 billion. Integrated operators, though much bigger, were reported as intending to invest $4.18 million. In thin flat-rolled products, the prospects for a continuing further severe thinning of the ranks were obvious. In detail, the consequences, would, as always, depend on the vagaries of innumerable company fortunes and the quality of board decisions.

NOTES

1. J. R. Miller, "Steel mini-mills," *Scientific American* 5 May 1984, 29–35.

2. T. Allen, formerly of J&L, in conversation with author October 1999.

3. H. Mueller, Obituary of K. Iverson, *Steel Times International* June 2002, 40.

4. H. Paxton, "Future Ferrous Technologies" 1991, 5.

5. *New Steel* January 1994, 56.

6. *New Steel* October 1993, 8; *Iron and Steel Engineer* February 1996, D1; Steel Manufacturers Association, website, 20 October, 1998.

7. *Iron and Steel Engineer* February 1996, D24; K. E. Busse and D. F. Barnette to *Steel Success Strategies Conference*, June 2000.

8. *Iron and Steel Engineer* February 1995, D1.

9. R. M. Cyert and R. J. Fruchan, *Sloan Steel Industrial Competitiveness Study*, Carnegie Mellon University, 1996.

10. *New Steel* August 1999, "The changing dynamics of steel making."

11. *New Steel* March 2000, 54; WSD web digest September 2000.

12. *New Steel* March 2000, 52.

13. WSD web digest and *New Steel* April 1999.

14. *AMM*/WSD, *Steel Success Strategies Conference* June 2000; *New Steel* September 1999.

15. *S.S.S Conference,* op cit.

16. *S.S.S Conference,* op cit.

Chapter Sixteen

Into a New Millennium

In the first few years of the twenty-first century the American steel industry and its strip mill sector, were in ferment. As an authoritative industrial journal put it as early as 2005, "Mergers, acquisitions, bankruptcies and restructurings have rapidly changed the steel industry's structure, coinciding with a high pace of new developments and several historical and unforeseen events."[1] The impact of change was complicated, colorful, profitable to some and generally painful for most. Throughout this trying period the operators of strip mills showed a range of characteristics—tenacity on the part of many, ingenuity by some, a readiness to capitulate in a few cases. The causes and results of the difficulties of these years were various.

There was first an increasing influence from the wider world of steel. Here one measure of the momentous changes of the new age may be sensed in two facts. As late as 1990, when world steel production was 770 million tons. China made only 66 million metric tons and the United States 90 million. In 2004, when global steel output for the first time exceeded one billion tons, 25.8 percent of that output came from China. In 2015, output of crude steel was 79 million tons in the United States and 804 million in China.

In steel circles in the United States there persisted an idea that the increasing competition from overseas was unfair. For example, in a *Wall Street Journal* interview, Thomas Usher, chairman of US Steel, characterized his industry as "the victim of a perverse system where foreign countries have used market barriers and subsidies to encourage excess capacity and production, which is then dumped in the United States—the largest and most open market in the world." Yet an early decision of the first Bush administration to impose swinging additional tariffs on imported steel was an example of active support from central government. The traditional commercial objection to such support still applied: if subsidies encourage production even when it is in excess of demand, protection permits inefficient firms to survive. Apart from opposition by the international steel community, which, with accompanying retaliatory action in other spheres of trade, eventually secured the relaxation of the Bush protective tariffs, practical considerations at home limited how much imports could be restricted. The reasons for this may be summed up in the estimates made in a study by the Consuming Industries Trade Action Coalition that a 40 percent duty on imports would cause approximately eight other jobs to be lost for every steel job preserved.[2]

A consideration affecting the distribution of the industry, and especially of its EAF sector, has been the existence and distribution of Right-to-Work laws. As in other long-established nations, both central and regional government have sometimes intervened, helping to preserve facilities which on any balanced view of trends would be classed as uncompetitive, and whose rescue only worsens future prospects for the industry as a whole. In 2001, American steelmakers received support of over $1 billion in the form of subsidies through US loan guarantees. Wheeling-Pittsburgh filed for Chapter 11 protection in November 2000. In March 2003, it was granted a loan of $250 million from the Federal Emergency Steel Loan Guarantee Board which helped it emerge from bankruptcy and revamp its production. Another instance of public assistance at a lower level of government was a 2004 decision by the state of Kentucky to contribute to a $65 million modernization of the Ashland works, which no longer operated its own strip mill, but instead supplied slabs to Middletown, from which in return it obtained part of the cold rolled coil for a hot dip galvanizing plant.[3]

The operating environment for the industry was even more volatile than in recent decades, costs and prices oscillating sharply. July 2001 prices for hot and cold rolled sheets were 20 percent and 21 percent respectively below those of a year before. Four years later, a steel industry consultant forecast that the 2007 average price for hot rolled coiled band would be as much as $650/ton.[4] Directly or indirectly, such price swings are increasingly related to situations in the wider world. Periods of increased pressure from foreign steel cause falls in domestic production and prices. From summer 2004, general activity in the US industry declined—in the year to June 2005 the number of blast furnaces at work nationwide fell from 32 to 23. There were reductions in carbon steel flat rolled products prices from early 2005, but in the first four months of that year imports of steel (of all types) were 22.8 percent higher than in the same period of 2004.[5] Scrap prices at home rose to new heights as the boom in steel output worldwide pushed up demand. Early in 2002, scrap in the United States was about $70/ton; two years later it reached $220. Scrap bundles sold for over $300/ton early in 2005, afterward fell to a low of $150 but by August had rebounded to $210/ton.[6] Nowadays major producers introduce, increase, decrease or remove surcharges to compensate for such price changes, but this degree of volatility increases the general uncertainty. (There remain very marked regional differences in scrap prices. For example, in June 2005, number 1 heavy melting scrap was $120/gross ton delivered in both Chicago and Pittsburgh, $115 in Cleveland, $110 in Detroit, but only $60–$62 in the Houston area and in the Carolinas as low as $52–$54. Around St. Louis, it was $35/ton.[7] The advantage provided by low regional prices to footloose electrical steelmakers is obvious.)

Some measure of the consequences of this uncertain operating environment for the strip mill industry may be gained from the fact that in a few years a number of prominent companies ceased to exist as independent steelmakers. Their place was taken by new groups. Of the latter the leader was Mittal Steel, which by 2005 controlled seven strip mills with a total nominal capacity of 22 million tons. It became part of an even larger international group, Arcelor-Mittal. The United States Steel Corporation in 2003 absorbed the operations of National Steel Corporation, and by 2005 it had about 18 million tons of strip mill capacity in five works. AK Steel, the third company, operated three strip mills, only one of them part of a major, fully integrated plant. Its strip mill capacity was about 6.5 million tons. Six other integrated operations together aggregated 11.3 million tons.

A number of strip mills have been closed and their equipment sold. (One company—Casey—has advertised as "the recognized experts in the sale, dismantling and selling of hot strip mills." At one time it had on its books a two-stand hot mill that depending on product mix could make 150,000–750,000 tons a year, a semicontinuous mill capable of more than 1 million tons and a 4-million-ton mill.)[8] Of the 29 hot mills existing in 1989, seven were out of production by 2003. Over the same period, the EAF sector installed nine strip mills. Effective capacity at conventional strip mills fell by some eight million tons, and 18 million tons was added by mini mills. In a volatile market, such a large addition of new plant, made when many old-style operations have spent heavily to maintain their positions and with imports at high, if fluctuating, levels, complicates the situation. It may be helpful to divide the present industry into five sectors: the recent failures; cases in which exceptional circumstances seem likely to preserve the plant; mills which are geographically peripheral; and a fourth and major group occupying the core region of the industry. The final section is the mini-mill sector.

In 2002, Wilbur Ross bought the steel assets of LTV. (By not buying LTV itself, liability to pay pensions to retired former employees was avoided!) After being idle for years, the 80-inch mill installed in the 1960s by J&L on the west side of the Cuyahoga in Cleveland was sold to Chinese interests in 2003 by its new owners, by this time operating as the International Steel

Group (ISG). Blast furnace, basic oxygen converter and slab casting facilities remained, capable of operation with the adjacent Cleveland East works (the former Republic Steel plant). Rouge Steel, operating under Chapter 11 bankruptcy, lost some $360 million and in October 2003 was sold to Severstal, a Russian group. That year it was rumored that US Steel was interested in acquiring Rouge, but it did not go ahead. Not far from the Rouge plant, there was at that time another instance of the mortality of even favorably located plants. McLouth Steel, first bankrupt in 1981, again filed for bankruptcy in 1996. After being idle for years, in 2005 its Trenton melting shop followed its two blast furnaces into demolition.

Integrated works in the peripheral regions of the national economy suffered severely. The owners of the strip mill at Geneva filed for bankruptcy in 1999, operations failed in 2002 and two years later the mill equipment was sold for $35–$40 million to Qingdao Iron and Steel of China. From August 2000, Gulf States Steel began to close its Gadsden plant; by mid-2005 the 900,000 ton hot strip mill was for sale.

Although it is noteworthy that the Cleveland West and the Rouge works were the first of the second-generation strip mills to close, the next category of mills, though smaller in total tonnage, is in some ways more interesting and more hopeful. The Acme strip mill at Riverdale, Chicago, closed in 2001, but the next year it was acquired by ISG, whose chairman made this revealing remark: "These are facilities that cost something approaching $350 million to build. We're buying them for $65 million." Late in 2004, Acme was brought into the Mittal group. Riverdale had a compact strip mill like some of the mini mills but also supporting blast furnace and BOF capacity. Its special asset was an ability to roll exceptionally wide thin-gauge band. The other large specialist works, the Warren strip mill, the first of the second-generation mills, was sold by a bankrupt LTV in 1988 to Warren Consolidated Industries. $350 million was spent by 2000 to upgrade it as a "niche-oriented integrated producer of value-added, custom steel products." The company advertised 185 grades of product, including high carbon, alloy, high-strength silicon, electrical and galvanized sheet, often in short runs, but made in an "operationally flexible" mill, the emphasis being said to be on "customer specific quality requirements and delivery performance." By 2014, and now owned by bankrupt RG Steel, the blast furnace, caster and cold mills were being demolished, and it was feared the hot mill would never work again. Three other specialist producers, the Butler and Mansfield works of AK Steel and Allegheny-Ludlum operations at Brackenridge, each had hot strip mill capacity of 600,000–700,000 tons, making special flat rolled products, stainless, electrical grade, etc. In a more and more sophisticated industrial economy, such operations seem likely to possess greater long-term survival value than mills making carbon steel grades.

Three strip mills were in marginal locations for carbon steel thin flat rolled products. Of these by far the most important was Sparrows Point. Its 3 million ton capacity, little more than half the rated total in its peak years, served a strip mill originally installed in 1947, but which had undergone major modernizations in the mid-1960s and the 1990s. For iron production and steel manufacture, Sparrows Point remained a highly favored location, and its strip mill had regional outlets for sheet, strip and tinplate. Even so, it illustrated the dangers of excellent natural advantages cancelled out by the unfortunate course of events. In March 2008 it was sold by Mittal Steel to Severstal for $810 million. Four years later it passed (along with other smaller operations in West Virginia and Ohio) to the Renco group for $1.2 billion. It filed for bankruptcy in May 2012, steelmaking ceased, and in August the plant was purchased for $72 million by Hilco Global, a business ominously (if colorfully) described as "a financial services firm with expertise in monetizing distressed assets." Two years later, the 3,000-acre site had passed to Sparrows Point Terminal LCC, and the steel plant had already been largely demolished. Its site was now a scene of extraordinary dereliction, as described by the *Baltimore Sun*: "The Sparrows Point peninsula today looks post-apocalyptic, a landscape of gray and brown, run through with puddles and cluttered with piles of scrap metal, giant nests of rusted rebar, abandoned trucks and other leftovers from more than a century of the site's life as a steel mill."[9] It was difficult to credit the fact that for a brief period some fifty years ago it was the world's biggest steelworks and characterized by the Bethlehem Steel chairman as "the greatest show on earth."

Fairfield was, after the closure of the former Republic Steel mill at Gadsden, the only integrated operation in the Southeast. On the west coast the Fontana operations of California Industries seemed less secure. Its 86-inch strip mill, after four important modernizations since it was installed almost fifty years before, continued to produce hot and cold rolled coil and galva-

Table 16.1. Type and geographical classification of hot strip mill capacity, 2003–2004 (million tons)

Traditional mills	
Core area	4.2
Special products	4.5
Peripheral	6.9
Recently failed	3.7
Total	59.3
EAF or "mini" mills	18.0
Grand Total	77.3

nized products, but had long been rolling them from purchased slabs. For many years it was highly profitable, but in 2003 it was caught in a serious cost/price squeeze, the cost of slab costs rising faster than the prices it could command for its output.[10] It benefits from the protection of distance from domestic land-bound rivals, but is exposed to competition from larger integrated overseas operators possessing more modern mills.

The 12 plants in the "core area" had in 2003 a strip mill capacity more than twice that of the 12 in the three classes considered so far. They were parts of fully integrated works, on average were much bigger and rolled wider strip. In short, they were the leading contenders to meet the requirements of the automobile industry and other large-scale users of high-quality, carbon steel sheet, wide strip and coil. Some 18 million tons of the 44 million tons of strip mill capacity in this class was in the Chicago area. US Steel then controlled 6.3 million tons of this at Gary, and Mittal Steel 11.8 million tons, divided between three works, Burns Harbor, and the former Inland and Youngstown Sheet and Tube (later LTV) works at East Chicago/Indiana Harbor. If a decline came in the activity of this region, it might be that one of the last two would be idled. The major Portage cold reduction mill and the similar operation built in the 1960s by J&L at Hennepin remained, but company realignments in the early 21st century changed possible supply patterns.

In March 2002, National Steel, for almost all its seventy years a highly profitable company, filed for Chapter 11 protection. It continued for a time to operate its plants as "debtor in possession," but in spring 2003 sold the iron, steel and raw material operations to the United States Steel Corporation for what seemed a very modest price of $1.05 billion. The National subsidiary, Midwest Steel, which had finished much of the material rolled on the hot strip mills at its two major works in Detroit and at Granite City, now became linked with Gary, located only about 10 miles from the Portage cold reduction mills. Longer-term prospects at Granite City seemed likely to suffer from these changes. At the same time, formation of the Arcelor-Mittal group meant the Hennepin cold mill complex, originally designed to be supplied with hot rolled coil from Cleveland (and other J&L works), now had two nearer sources of supply in the hot strip mills of Indiana Harbor and Burns Harbor.

The former Republic, later LTV, strip mill in Cleveland had a large market area of its own and survived. Weirton, though landlocked, and having suffered many troubles, continued for a time as Arcelor-Mittal's main source of medium-width coils for tinplate production. Middletown, the main AK Steel mill, and at 5.3 million tons, second in size only to Gary in rated strip mill capacity, also continued.

There remained serious questions concerning the longer-term prospects of three plants in the northeastern interior—Sharon, Steubenville, and the only surviving Pittsburgh mill. Duferco took over the Farrell works of the former Sharon Steel Corporation. This was an integrated operation, but the smallest of all the carbon grade strip mills, and with the added disadvantage of being a wholly land-locked operation, lacking even river transport facilities. The most interesting question of all was the future of the Irvin works. Built in 1938, it was, excepting Fairfield and Weirton, the oldest of the major strip mills—although "modernized" in 1995 and 1998. Following the acquisition of National Steel, United States Steel controlled three other hot strip mills in the manufacturing belt, each a second-generation mill and bigger than Irvin. Gary and Ecorse are better located in relation to raw material supply and outlets; the third, Granite City, was in these respects no better than Irvin. If USS found it necessary to further rationalize its steel mills in this region, Irvin works and the iron and steel units at Edgar Thomson that supplied it seemed prime candidates for closure. Should this happen, the history of bulk iron and steel manufacture in the Pittsburgh area would at last come to its end.

Many major companies invested heavily in a struggle to maintain their place in the trade. The experience of Wheeling-Pittsburgh provided an interesting example, which may merit fuller examination. This merger was created as a result of the financial burdens resulting from the investment in the second-generation mill at Steubenville. There were further serious problems in the 1970s and 1980s. Company operations were divided between seven works. Although until August 2003 it operated under Chapter 11 bankruptcy protection, Wheeling-Pittsburgh began to modernize its plant. The program was comprehensive. In 2000 new segments were added to the twin strand continuous casting unit to increase its productivity. It continued to operate BOF capacity but late in 2004 commissioned an electric furnace plant which was reported to rapidly prove itself in terms of high productivity. It was designed to use either a 100 percent scrap burden—converting the mill in this respect into an operation like the mini mills—or a mixture of molten iron and scrap. In this way it was hoped the company's steelmaking could be well-placed to respond to any changes in material prices. Although its new EAF plant was designed to enable closure of one of the two blast furnaces, investment in iron making was not neglected. In spring 2004 a new metallic heat exchanger was brought into use at number 1 blast furnace; by preheating gas, it was hoped that coke consumption might be cut by 3,000 tons a month. The first equipment of its kind in the industry, this cost $4 million, but it was claimed would pay for itself in savings on the coke account within two months! In spring 2005, the company announced the installation of automatic roll changers on its strip mill, estimating this would reduce the time needed for each change of rolls from as much as 40 to as little as eight to ten minutes. On average rolls were changed four to six times in a 24-hour running time. Costed at

$14.5 million, this equipment was reported to increase mill effective annual capacity from 2.8 to 3.2 million tons and improve productivity. In short, this company's experience seemed to show something of the innumerable ways in which costs could be reduced and existing strip mills helped to survive.[11]

Inevitably the uncertainty and commercial problems in the industry were accompanied by yet another social crisis. There were sharp cuts in workforces. Pay and benefits were renegotiated and usually savagely reduced. Though the circumstances are radically different, the elements are not dissimilar from those in earlier transformations in the sheet industry. Reductions in employment and new, less favorable conditions of work for those who remain are regarded by the companies, and to some extent by unions, as a necessary price for the preservation of livelihoods. Essential parts of Wheeling-Pittsburgh's program for survival were payroll cuts and many changes in employee and retirement benefits.[12] Early in 2004, facing up to the fact that from being a leader in the industry it was now suffering sharply reduced profit margins (in the second quarter of 2005 it earned only 8 cents a share whereas USS earned $1.62) AK Steel notified unions representing 3,100 hourly employees that it was suspending a contract in force since 1995 which had guaranteed a minimum number for employment at Middletown, a works which had for some time been regarded as the most productive of the nation's integrated operations. By August, men were being laid off there on the grounds that its workforce was too large and that it was at about a $30/ton disadvantage compared with some other integrated works.[13] At Weirton Steel, the employee share of the common stock had fallen to 25 percent by 1996, at which time the company employed about 5,600. In spring 2003, after losing $750 million over five years, Weirton filed for Chapter 11 protection, listing assets worth $654 million against liabilities of $1.4 billion. An important aid to its struggle to survive involved obtaining $278 million in concessions, partly from workers—who accepted a 5 percent pay cut and a freeze in benefits including pensions—partly from suppliers and lenders. By the time of the failure, employment was down to 3,500. In May 2004, when ISG group bought Weirton for $255 million, it was revealed that a restructuring plan provided for a halving of management posts to 200 and a 1,000-person cut in the mill workforce. At the same time Weirton announced plans to cancel health benefits for almost 10,000 retired workers.[14] WCI provided a variant on the same theme. Having filed for Chapter 11 protection in September 2003, four months later it announced measures designed to save $3 million a year, including a 10 percent fall in salaried workers and a decrease in company contributions to their savings plan. In August 2005 it announced a "Plan of Reorganization."[15]

Another, very different and less controversial means of improving the prospects of the strip mills has been product improvement, enabling companies to command greater consumer preference and a higher price. The vital importance of this was made clear in summer 2001, when General Motors unexpectedly revealed it was dropping LTV Steel from its list of suppliers, and that its purchase agents "intend to work with a smaller number of steel makers that can consistently drive high levels of quality, delivery and innovation at competitive cost levels." Until then, LTV had supplied GM with about 500,000 tons of steel a year.[16] Loss of this business was followed early in December by closure of the LTV Cleveland operations.

Upgrading the finished product was another way of meeting increasing competition. It began with the molten metal. The changes are summed up under the general heading of "ladle metallurgy," involving removal from the molten steel of undesirable nonmetallic elements. This further refining of the metal from either BOF or electric arc furnaces may involve vacuum degassing, bringing about removal of harmful gas inclusions, especially hydrogen. At a later stage of manufacturing, more of the firms have installed walking beam slab heating furnaces—so-called because the slab passes through the furnace on a moving beam, rather than, as in the past, being simply pushed through. More complex and costly than the old-type furnaces, the new arrangements eliminate sticking and skid marks on the steel and provide for more even heating. Decarburization is a means of removing the carbon-depleted surface of the steel, thereby making a more uniform product. Finally, ever more precise controls in the strip mill and in heat treatment of hot rolled coils can upgrade finishing operations. All these elaborate and costly innovations have helped satisfy more exacting market specifications, and meeting challengers from overseas and from the expanding EAF strip mill sector at home.

In 1989 the rated annual capacity of American strip mills was almost 68 million tons. Not until August that year was the first mini-mill strip mill commissioned. By 2004/5, national strip mill capacity had increased to 74 million tons, but the capacity of conventional strip mills had fallen by nearly 12 million tons and there was now some 18 million tons of capacity at mini mills. The number of conventional mills fell from 29 to 20; the number of EAF mills increased from none to nine. Yet as the long-established sector of the industry suffered trauma and struggled to survive, mini mills too have faced serious difficulties and suffered casualties. As long anticipated, the operators of electric furnaces have been penalized by the oscillations in scrap prices. As they rose, both mini mills and integrated companies imposed "surcharges" on published prices, but inevitably greater dependence on this material meant the former had to add more. For example, early in 2004, when Wheeling-Pittsburgh increased its surcharge from $30–$42/ton, Nucor and Steel Dynamics were moving on from $60 to $100. Both of the latter companies invested in "scrap substitutes," and by early 2005 the president of Steel Dynamics mentioned that his company was thinking of acquiring its own scrapyards. By 2005 Nucor had a more than one fifth interest in a charcoal iron project in Para State, Brazil, the whole of whose 360,000 tons of pig iron it was expected to consume.[17] Yet it seemed that this would be only of marginal importance for a company which in 2004 used 17 million tons of scrap; Nucor 2003 raw steel production was 17.4 million tons. Eleven years later, Nucor use of scrap remained about 17 million tons, but steel output was almost 29 million tons. It had achieved a major widening of its raw material base.

In that sector of the mini mill industry concerned with thin flat-rolled products there have been some disappointing performances. Trico Steel, which began operations in Decatur, Alabama, in 1997, had a hot strip mill capacity of 2.2 million tons. The works closed and the company became bankrupt in March 2001. Sixteen months later, it was purchased by Nucor. The 800,000 ton electric furnace capacity installed at Portage, Indiana, to support the high-quality steel coil operations there of Beta Steel stood idle by spring 2004.[18] Nor did bigger EAF companies escape. By 2003/4 Nucor was, at this time only for a short period, the largest steelmaker in the United States, with 2004 net sales of $11.3 billion. But profitability suffered in the cost/price squeeze of the first years of the new millennium. For instance, in 2003 Nucor's scrap and "scrap substitute" costs increased by $37/ton and energy costs by $3/ton. Together with depressed levels of activity in its markets, these higher energy and material costs led to a 52.1 percent fall in the company's net income in the fourth quarter of 2003 as compared with the same quarter a year earlier despite a 28.3 percent increase in sales.[19]

Generally, mini mills improved their standing in terms of product quality, took steps to reduce the uncertainties in their raw material supply situation, and continued to benefit from lower standing charges as well as from their well-established advantages of location, and as a result of their favorable management and workforce conditions. They continued to pioneer technical improvements. A key sector of the latter has been the way in which BHP and, following it, Castrip at Charlotte, North Carolina, worked out a practical process for direct casting of molten steel into both the shape and the thickness of the final product without further rolling. In May 2002 Nucor began "Castrip"-type operations at Crawfordsville. The process makes a smaller plant practicable, seems likely to reduce capital outlay and by economizing in energy cuts operating costs—as Nucor put, it such a "micro mill" could be constructed in one sixth of the area a mini mill would occupy and at only on tenth of the outlay on an integrated works. It claimed it would consume 95 percent less energy than the integrated plant. Adding to these assets, it has been suggested that the Castrip process produces a product superior to ordinary grades of hot band. By early 2005 Nucor's Castrip process was already reckoned to have achieved commercial viability.[20]

Over the next few years, the EAF strip mill sector continued to grow, and its leading companies increased their dominance within it. In each of the years from 2011 to 2016, the crude steel production of United States Steel fell; the only exception was the calendar year 2015, when production by Nucor increased. In 2014, Nucor acquired Gallatin Steel of Ghent, Kentucky, and Steel Dynamics bought the steel and strip mill operation constructed by Severstal at Columbus, Missouri. But by far the largest of the single EAF steel and strip mill operations has been built by overseas investors in the lower part of the Mississipi Valley in the near vicinity of Mobile. Between 2007 and 2010, Thyssen-Krupp built a plant with a hot strip mill capacity of 5.3 million tons at Calvert, 35 miles north of Mobile at a new river terminal on the west side of the Tombigbee River and near the head of a deep bay opening into the Gulf of Mexico. This plant largely finishes its thin flat-rolled sheet as galvanized material and serves automotive, construction, appliance and pipe and tube industries. Early in 2014, Calvert was bought by Arcelor-Mittal in an equal-share partnership with Nippon steel and Sumitimo Metal Corporation. Claimed by Lakshmi Mittal as "the most modern steel finishing facility in the world," it was announced in 2015 that it would receive a further $88 million investment.[21]

The shape of the new steel industry and of its strip mill sector has decisively changed. There has been a sharp reduction in dependence on blast furnaces—pig iron production fell from 42.1 million metric tons in 2001 to only 26.8 million nine years later—a 36 percent fall over a period in which raw steel output declined less than 11 percent. The trend to electrical steel continued. In 1991 electric furnaces made only 38.4 percent of US steel; by 2000 their share was 47.0 percent and in 2010 61.3 percent. This shift marked the further advance of the EAF sector and therefore of the newer as compared with the long-established steel companies. By 2014 Nucor crude steel capacity ranked it as the biggest steel company in the Americas. Its growth and that of other EAF producers had occurred despite what might seem to have been adverse trends. Scrap prices have continued to be, in the words of one enquiry "notoriously volatile," and in spring 2011 an OECD study projected that over the next 4 years increases in world steel capacity would put more pressure on its supply.[22] The EAF sector continued to respond to this challenge in two main ways: by better organizing scrap supplies and by developing alternative sources of metal.

In 2007, Nucor was reckoned to "recycle" almost 10 million cars in its production processes. The following year, it spent $1.4 to acquire the David J. Joseph Company, one of the nation's largest scrap-processing firms. In

Table 16.2. Steel production by leading companies, 1994, 2002, and 2010 (million tons)

	1994	*2002*	*2010*
United States Steel	11.5	14.4	22.3
Nucor	6.6	12.4	18.3
AK Steel	3.7	5.3	?

2007, Steel Dynamics gained control of Omni Source Corporation, which had over 80 major recycling operations in the Midwest and the Southeast. A few years later, Omni was supplying about half the scrap used by the five Steel Dynamics works.

The other line of advance has been to secure control of iron ore deposits and develop more efficient means than the blast furnace of converting ore into feed for electric arc furnaces. Nucor seems to have been disappointed in the efforts it had made some years ago to develop low-cost sources of iron in Brazil and in Australia, but in 2011 it invested $750 million in a 3-million-ton direct reduction (CDRI) plant using ore pellets and natural gas at St. James Parish, Louisiana. This required a new river port, designed to handle both oceangoing ships and ore- and coal-carrying barges. By the end of 2013, 2.5 million tons of annual capacity was in production there. Steel Dynamics remained above all dependent on scrap—supplying all but 8 percent of the 6.3 million tons of metallic materials it consumed in 2013. But it too acted in direct reduction, in 2007 acquiring or leasing a total of almost 6,000 acres on the Mesabi ore range, and two years later at Hoyt Lake, Minnesota, a former LTV operation, and having Kobe Steel as a minority partner, began a project using railroad-delivered Appalachian coal to make high-purity iron "nuggets" for rail transfer to its steel operations.[23] By further developments along similar lines, EAF works may be able to meet the challenge of tighter, costlier scrap supplies.

The latest and major EAF project, Big River Steel, was begun in 2013 by a group headed by John Correnti, who over a long career had worked for US Steel and Nucor. He died in summer 2015 while the new works at Osceola on the Mississippi in Arkansas was still under construction. Its 1.6-million-ton capacity will bring the capacity of EAF-based strip mills to almost 30 million tons. None of this capacity existed before Crawsfordsville came into production some 26 years ago. In this period the "traditional" strip mill sector has fared badly, the rated capacity of its strip mills falling by almost 15 million tons. The relative standing of the two sectors in production is less clear, but it is certainly even more unfavorable to the old-style production cycle. There seems to be no reason to doubt that these trends will continue.

Table 16.3. EAF strip mills, 2003 and 2015 (capacity in million net tons)

	2003	*2015*
Nucor		
Hickman, Ark	2.0	2.5
Berkely, S.C	2.5	2.5
Decatur, Alab (ex Trico)	2.2	.5
Crawsfordsville, Ind	2.0	2.1
Ghent, Ky (ex Gallatin)	—	1.7
Steel Dynamics		
Butler, Ind	2.4	3.0
Columbus, Mo (ex Severstal)	—	3.4
NMLK, Portage, Ind (ex Beta)	?	1.1
North Star/BHP, Delta, Ohio	3.0	3.0
AM/NS, Calvert, Alab	—	6.0
Gallatin Steel, Ghent, Ky	1.5	—
Total EAF Mills	15.6	27.8

Based on AIST Hot Strip Mill Roundup 2003, 2015

Some strip mill operations which seemed able to hold on in the early years of the new millennium later lost out. By 2008, Weirton, which once had 12,000 workers, already employed only 1,200. The strip mill there rolled its last coil in December 2007. After that, only its tin mill remained at work. The bankruptcy of RG Steel in spring 2011 caused the failure of two plants. The strip mill at Warren, Ohio, closed, and by summer 2014 its cold mill, con-caster and blast furnaces were scheduled for demolition. The hot mill was reckoned unlikely to find a buyer and was demolished in 2014. Sparrows Point was a much bigger RG Steel casualty.[24]

In summer 2014, it was announced that Severstal would sell two works—its Columbus, Missouri, EAF steelworks and compact strip mill for $1.63 million to Steel Dynamics, and the Dearborn strip mill, which Ford had built, to AK Steel for $700,000. By this time two strip mills continued to depend on purchased slabs—the Fontana works of California Steel Industries, and the Farrell, Pennsylvania, mill acquired in 2011 by NLMK. The high hopes once entertained for reconstructed Wheeling-Pittsburgh operations were dashed. Its new blast furnace and electric furnace failed to meet their targets; the electric arc furnaces installed at Mingo Junction at a cost of $115 million were sold off for $20 million. The company was beset with bankruptcy hearings and entered on the sad process of passing from one speculative purchaser to another—to Esmark in November 2007, Severstal 10 months later, and in 2011 to RG Steel, whose 2012 application for Chapter 11 bankruptcy brought its closure. The Martins Ferry division was sold to a local businessman, and Steubenville, the key operation, went to a local "metal recycler." By 2014, the only former Wheeling-Pittsburgh facility still at work was the Follansbee coke works.

The whole of the rest of the strip mill industry is now owned either by EAF companies or by one or other of the three surviving integrated groups. AK Steel has concentrated operations at Middletown, which it claims as the most productive integrated works in the United States. For many years, Ashland retained its blast furnace and basic oxygen steel operations but the hard trade conditions of 2015 at last caused its collapse. At the end of the year, it was announced that Ashland iron making and steel production would cease. Possibly fewer than 200 of its workforce would survive in hot-dip galvanizing operations. Mansfield works has been turned into what is effectively a minimill operation, with electric arc furnaces, modern "ladle technology" and thin slab casting serving its 6-stand hot strip mill. It specializes in stainless steel, for which it has a 600,000 ton annual capacity. Butler produces stainless, electrical, and carbon grades.

There remain two very large, but contrasting groups. The North American division of the worldwide combine Arcelor-Mittal is heavily concentrated in the Chicago area. At Burns Harbor, and in the surviving parts of the old YS&T works and the old Inland works at Indiana Harbor, it controls a steel, capacity of about 15 million tons and three hot strip mills. Riverdale continues to be supplied by its BOF steelworks, which uses as much as 1 million tons of hot metal carried from the blast furnaces at Indiana Harbor. The A-M Cleveland strip mill and integrated works survives. The 2013 rate of utiliza-

Table 16.4. Changes in the distribution of strip mill capacity, 1991–2015 (million net tons)

	1991	*2015*
"Conventional" mills		
Mid-Atlantic coast	4.8	—
North East interior	24.0	16.3
Lake Erie and Detroit	16.2	11.5
Lake Michigan	18.4	21.2
Southeast	1.9	2.0
West	3.3	3.0
TOTAL	68.6	54.0
EAF strip mills	—	29.4
GRAND TOTAL	68.6	83.4

Based on AISI statistics

tion of the capacity of the strip mills at the three AM complexes is thought-provoking—53.7 percent for Cleveland, 78.9 percent at Burns Harbor and 92.9 percent in Indiana Harbor.[25]

The United States Steel Corporation is bigger, and its operations are more widely spread. Until 2015, it had five strip mills. Two were in prime Great Lakes locations—Gary, with a raw steel capacity of 7.5 million tons, supplies its own 6.3-million-ton strip mill and sends hot coil to be further processed at other USS works in the area. The second-generation strip mill at the former Great Lakes Steel Ecorse works mostly serves the automobile industry. The three other strip mill locations proved less viable. At Fairfield, Alabama, steel capacity supported a major output of tubular products as well as the strip mill. Granite City, acquired with National Steel in 2003, had a hot strip mill which was modernized in 2009. The capacity of its steel operations was considerably less than that of its hot mill. Most intriguing of all is the outlook for Irvin works, on whose existence depended the last remnant of bulk iron and steel in the Pittsburgh area. The so-called Mon Valley complex also continued to supply cold rolled material to the Fairless galvanizing plant, the only operation surviving there.

In 2015, a fall in demand, the effects of a strong dollar and rapidly rising imports—which supplied 14 percent of the sheet steel market in 2012 but 22 percent two years later—meant that USS was able to operate only about 60 percent of its flat-rolled products capacity. Prices were falling. The US home price for hot rolled sheet was $684/ton in May 2014, but by the end of 2015 only about half as much. This situation caused a major reconstruction of the Corporation's strip mill operations. It began 2015 with five strip mills; a year later only three were at work. At Fairfield iron making, the steelworks, and hot and cold strip mills were permanently closed. Granite City steel and strip mill operations were closed "temporarily" from the end of the year with the laying off of 2,000 employees. (At the end of 2016, USS announced operations at Granite City would restart early in the next year.) These closures allowed the remaining three USS strip mills to raise their utilization rate and remain financially viable, but in the longer term Irvin at least seems vulnerable. By the middle of 2015, employment there was about 1,000 (as compared with 3,500 in 1981, when it had previously been "idled"). Another 2,500 worked in other USS Monongahela Valley works. The almost complete closure of the Fairless works, which in the early postwar years was believed to mark a breakthrough for iron- and steelmaking and strip mill operations on the east coast, the later collapse of the once massive Sparrows Point plant, and the uncertainties about the long-term future of the Irvin works, highlight the mutability of conditions. In 2014, when US production of hot rolled sheet, coil and strip totaled 20.7 million metric tons, United States Steel Corporation shipments of hot rolled and cold rolled sheets and coated sheet was 12.4 million net tons; two years later the respective figures were 18.7 and 9.2 million.

In 2004, 15 years after the first EAF/compact strip mill was built by Nucor, mini mills accounted for a little less than one quarter of nominal hot strip mill capacity. By 2015, their share was 34 percent and there could be little doubt that more of the capacity in their strip mills was active than in those at integrated works. By now it was obvious that compact mills and the companies which owned them would make further headway at the expense of the "old-style" mills. There seems no likelihood that thin flat-rolled products and the hot strip mills which produce them will be displaced from their central positions in the steel industry, but their future seems likely to be concentrated in the hands of either a small number of integrated companies mainly on the shores of the Great Lakes or mini-mill companies, more widely scattered through the nation. The revolution that began early in the twentieth century in the experiments or the pilot plants operated by Bray, Tytus, Naugle, and Townsend, and after that was taken up in bold commercial ventures by a handful of second-rank steel companies, has by no means run its course.

NOTES

1. *Iron and Steel Engineer*, "Developments in the North American iron and steel industry 2004."
2. A. Shavey, "The myths of the steel industry's arguments for increased government intervention," *The Heritage Foundation* 7 February 2000.
3. M. Applebaum 21 September 2005.
4. *Iron and Steel Engineer*, "Developments in the North American iron and steel industry 2004."
5. *Pittsburgh Post-Gazette* 14 August 2005; *AMM* 3 June 2005.
6. *CJAD Canada* 25 August 2005.
7. *AMM* 7 June 2005.
8. Website entry, "Wide continuous hot strip mill."
9. *Baltimore Sun* 22 June 2015.
10. *AMM* 15 January 2004.
11. Wheeling-Pittsburgh website.
12. *AMM* 14 May 2004.
13. *The Cincinnati Enquirer* 15 January 2004; *AMM* 16 August 2004.
14. *Pittsburgh Post-Gazette* 20 May 2004.
15. Warren Consolidated Industries website.
16. *Purchasing Magazine* online 9 August 2001.
17. *AMM* 4 January 2005, 11 February 2005.
18. *USWA* March 2004.
19. *AMM* 30 January 2004.

20. *World Steel Dynamics Conference* June 2000, panel iv, mini-mills; Nucor website.
21. *Birmingham Business Journal* 12 June 2015.
22. OECD May 2011.
23. Nucor and Steel Dynamics annual reports and websites.
24. Steel industry websites and *Youngstown Business Journal* 6 August 2014.
25. Arcelor-Mittal annual reports and website.

Appendix

The Old Sheet Steel Industry and Contrasting Perspectives on the Impact of the Strip Mill on Employment

I. LABOR IN THE TRADITIONAL HAND MILL INDUSTRY

In 1918, in the poem "Psalm of those who go forth before daylight," in the collection *Cornhuskers*, Carl Sandburg evoked the hard physical labor of those who worked in the sheet mills: "The rolling-mill men and the sheet-mill men are brothers of cinders; they empty cinders out of their shoes after the day's work; they ask their wives to fix burnt holes in the knees of their trousers; their necks and ears are covered with a smut; they scour their necks and ears; they are brothers of cinders."

II. PERSPECTIVES ON THE NEW TECHNOLOGY FROM AMONG INNOVATORS AND MANAGEMENT

In a discussion at the American Iron and Steel Institute in spring 1927, the vice president of the AS&T Company, C. W. Bennett, referred to recent developments in wide strip and sheet rolling "which mark distinct economic and humanitarian advances in steel mill operation." (1) Years later, John Tytus was given a silver-plated model of a rolling mill. He used it as a paperweight. As a historian of Armco recalled: "When men calling themselves 'sociologists' spoke of the continuous wide-strip rolling mill as 'a monster' because it threw men out of work, 'he would pick up this little silver mill and ask, quietly and politely: "Did you ever try your hand at the kind of work it ended?"'" (2) But, in the light of that perfectly reasonable answer, nobody seems to have asked him a supplementary question—why had his company for so long been content to extend this harsh labor?—or whether it would have persisted so long with an attempt to replace it if it had not above all been in the company's commercial interests to do so. Even so, the Bennett and Tytus comments focused one aspect of the social problem of the continuous mills.

III. THE EXPERIENCE OF THE ADVANCE OF THE NEW TECHNOLOGIES BY A FORMER MILL WORKER

In the late 1930s, when vigorous attempts were being made to unionize steel industry labor, the experiences of former sheet mill workers were collected. On 21 April 1940 Jos. J. Clair responded from Cleveland to Harold Ruttenberg, research director for the Steel Workers Organizing Committee (SWOC) Ruttenberg had written from Pittsburgh. Clair replied:

> Friend Harold,
>
> I received your letter and am only too glad to grant your request, so as you may continue your good work.
>
> Well, Harold, it's a long story filled with headaches and disappointments, but here goes.
>
> I first entered the steel mills in 1911 with the American Rolling Mill Company, Middletown, Ohio, as second catcher on hand sheet mills. I advanced from one job to another until I had learned them all. In 1921 I received my first rolling turn and continued in that capacity until 1926, when I left this company and accepted a position as roller on hand sheet mills with Otis Steel Company, Cleveland, Ohio, for which I am still working—not as a roller, but as a common laborer, which I consider one of the greatest setbacks in any man's life, because when a man worked as I have to reach the top, and when you have reached the top, only to be dumped and scrapped by the introduction of continuous machinery.
>
> Before the continuous mills were installed, I and my family were living very happily because we had a good standard of living, were properly schooling our children, enjoyed some of the privileges of life, and saving for the future. When I could no longer work in the mills, the future did not turn out as planned.
>
> In 1931 the Otis Steel Company placed into service the continuous mill, and then things started to happen, not all of a sudden, but gradually. At that time they operated ten hand sheet mills employing about 450 men. To make room for the continuous mill four hand-mills were taken off and dismantled. Work on the remaining mills was divided among all the men, cutting the time in half, but each week the orders for the hand mills declined more and more until at last we were only making a few junk orders now and then that the

company had to accept in order to get business for their strip mill. Pay after pay getting smaller, two or three days every two weeks, and most of the time no pay at all to take home.

The little nest egg that I and my wife had saved up for a rainy day was fast dwindling, for the rainy days never ceased, but I did not give up hopes, thinking that someday things would break–which they did, but for the worse. In December, 1937, the climax came. The hand sheet mills were discontinued and hundreds of men thrown completely out of work and to the mercy of the winds, but the company management was very considerate. They extended their sympathy for us, stating that it was impossible to operate hand mills and meet competition and promised that these men displaced by the use of the continuous mill would receive employment in other departments, but their promise never held.

We then placed our case before the SWOC in Cleveland. A committee was arranged and a meeting with the management was held, but with no results, only that they would take care of these men and give them some work. In 1938 all hand mill workers received through the mail an honorary discharge along with a record of service and reason for same (mill's discontinuance). Again the locals of SWOC committee met with the management but with the same results. This was the end of our rope so these men who toiled all of their lives in the mills were forced to seek a new field. You can imagine how this worked out for most of us, when you attain the age of forty years, no other industry wants you. Relief and WPA was their only salvation.

When I was working at my job as a roller, my average salary in the late years, after many cuts in wages, was $22.00 per day. Before the continuous mill it amounted to about $30.00 per day or about $6,000 per year with time off for slack periods. With no place to work and broke, I was forced to apply to WPA for work which I received at the $60.00 per month wages, which was impossible for me and my family to live on and keep house and make the payments on our home; so we were forced to sell the home, placed our furniture in storage where it now is, and take up quarters in a rooming house. The little cash we realized from the sale of our home also went, but we clung on never giving up hope, but the clouds continue to hang low.

In September, 1939, the company called some of us back and gave us jobs as laborers, the first we had worked in the mill since 1937. After working five weeks we were laid off or discharged because of work being completed. In March, 1940, we were called again, rehired, worked three days, and then fired again—same cause. 8 days we were rehired. So you see how they have taken care of us. Each time we get hired and fired, as our employment application only states temporary employment. Our work consists of the hardest and dirtiest kind, as we are only called back when something happens and needs quick repairs, which averages about two to four day's work. In fact we do not receive as much pay per month as when I was on WPA.

I do not know if they are holding a grudge against the sheet mill workers or not, but it sure appears that way.

Harold, there is a lot more that could be added to this story pertaining to the life of a sheet mill worker as I see it and hear it from those who have suffered through the use of continuous machinery. I know you are in position to obtain some of these facts from men in other mills, but when you see these men that you have worked with and know were good law-abiding citizens that were always respected for their ability to be self-supporting, being driven down and down until they did not care what next happened, their homes lost, families split up, and the whole reason is that the greed of our industrial leaders to fatten their purse and to meet competition through the installation of continuous machinery, I think it about time a halt is called to stop this inhuman treatment.

So, Harold, I want to congratulate and wish you success in the work you are now doing. As I have given you only a brief outline of what happened in the mill, if you desire more information, just write me or arrange a meeting, and I will try to give you any information you are looking for.

Thank you for the interest you have taken for the interest of the steel workers.

I remain
Sincerely yours,
Jos. J. Clair
9412 Willard Avenue
Cleveland, Ohio

Source: Ruttenberg papers; Pennsylvania State University

See also H. J. and S. Ruttenberg, "War and the Steel Ghost Towns," *Harper's Magazine*, January 1940: TNEC hearings; Testimony of Phillip Murray and others, April 12, 1940, pp. 147, 148, 151, 165; ibid., pp. 182, 183, in which it is calculated that between 1929 and 1939 the number of men already displaced from sheet and tin plate mills due to the introduction of the strip mill amounted to 38,470.

Bibliography

GOVERNMENT DOCUMENTS

United States Congress. 1912. House of Representatives. Hearings before the Committee on Investigation of U.S. Steel. (Stanley Committee) Washington, DC: Government Printing Office.

United States Congress. Transport Investigation and Research Board. 1945. Economics of Iron and Steel Transportation. Washington, DC: Government Printing Office.

United States Congress. 1950. House of Representatives Subcommittee on the Study of Monopoly Power. Washington, DC: Government Printing Office.

United States Congress. Hearings before the Sub Committee on Antitrust and Monopoly. 1957/58. Senate Judiciary Committee Washington, DC: Government Printing Office.

United States Congress. Hearings before the Joint Economic Committee on Steel Prices, Unit Costs, Profits and Foreign Competition. 1963. Washington, DC: Government Printing Office.

United States Department of Commerce. Historical Statistics of the United States. Annual. Washington, DC: Government Printing Office.

United States Federal Trade Commission. 1924. In the Matter of the United States Steel Corporation et al. Complaint, Findings and Order. Washington, DC: Government Printing Office.

United States Industrial Commission. 1899–1901. Reports. 13 vols. Washington, DC: Government Printing Office.

United States Temporary National Economic Committee 1939/1940. Investigation of Concentration of Economic Power. Washington, DC: Government Printing Office.

BOOKS AND ARTICLES

Adams, A. 1956. *Report on Pittsburgh Steel Company to New York Society of Security Analysts.*

Adams, W. (ed.). 1950 and 1992. *The Structure of American Industry*. New York: Macmillan.

AISA. 1866–1912. *Bulletin* (BAISA), AISI *Yearbook.*

AISA and later AISI. Occasional. *Iron and Steel Works Directory of the United States and Canada.* New York: AISI.

Aldefer, E. B., and Michl, H. E. 1942 and 1957. *The Economics of American Industry.* New York: McGraw-Hill.

Armstrong, G.S. and Co. 1952. *An Engineering Interpretation of the Economic and Financial Aspects of American Industry. X The Iron and Steel Industry.*

Association of Iron and Steel Engineers. 1941. *The Modern Strip Mill.* Pittsburgh: AISE.

Association of Iron and Steel Engineers. 1970. *The Hot Strip Mill: Generation II.* Pittsburgh: AISE.

Association of Iron and Steel Engineers. Annual. *Developments in the North American Iron and Steel industry in . . .* annually in *The Iron and Steel Engineer.*

Auerbach, P. 1988. Competition: *The Economics of Industrial Change*. Oxford: Basil Blackwell.

Aylen, J. 2010. "Open versus closed innovation: Development of the wide strip mill for steel in the USA during the 1920s." *R&D Management* 40(2010).

Badlam, S. "Evolution of the Wide Strip Mill." AISI reprinted *Iron Age* 3 November 1927, pp. 1232–1235.

Berdis, A. J. 1951. *Development of the Fairless Works of United States Steel* 28 September.

Biggert, F. C. 1927. "Developments in 4-High Rolling Mills," Iron and Steel Section. *TAIME* vol. 49–50 1927–1928.

B.I.S.R.A. 1953. "The Fairless Works." London: British Iron and Steel Research Association.

Borrus, M. 1983. "The Politics of Competitive Economies in the United States Steel Industry." Chapter 2 in J. Zysman and L. Tyson. *American Industry in International Competition: Government Policies and Corporate Strategies*. Ithaca: Cornell.

Borth, C. 1941. *True Steel: The Story of George Matthew Verity and His Associates.* Indianapolis: Bobbs-Merrill Company.

Brody, D. 1960. *Steelworkers in America: The Nonunion Era.* Cambridge, MA: Harvard.

Broude, H. W. 1963. *Steel Decisions and the National Economy*, New Haven: Yale.

Burns, A. R. 1936. *Decline of Competition.* New York: McGraw Hill.

Cammens, L 1928. "Sheet Rolling." *TAIME* Vol 49–50 1927–28 I&S Section, pp 1–14.

Cartwright, W. F. 1945. "Modern American Steelworks Practice." *Transactions of the South Wales Institute of Engineers* 15 November.

Casson, H. N. 1907. *The Romance of Steel.* New York: A. S. Barnes.

Clark, V. S. 1929. *History of Manufactures in the United States*. 3 vols. New York: McGraw-Hill.

Commons, J. R. 1929. "Wages: statistics of the USA," *Encyclopedia Britannica*.

Cordero, H. G. 1957. *Iron and Steel Works of the World*. 2nd edition. London: Quinn Press.

Cordero, R. 1974. *Iron and Steel Works of the World*. 6th edition. London: Metal Bulletin.

Craig, R. S. 1966. *A History of the United States Steel Corporation and Predecessor Companies: A Case Study of American Business*. Unpublished: US Steel.

Cronemeyer, W. C. 1930. "The Development of the Tinplate Industry," *West Pennsylvania Historical Magazine* pp. 22–54, 123–135.

Crout, G. C., and Vorhis, W. D. "John Butler Tytus: Inventor of the Continuous Steel Mill." *Ohio History Journal*.

Cyclopedia of American Biography. Various dates.

Cyert, R. M., and Fruchin, R. J. 1996. *Sloan Steel Industry Competitiveness Study*. Pittsburgh: CMU.

Daugherty, C. R., de Chazeau, M. G., and Stratton, S. S. 1937. *Economics of the Iron and Steel Industry*. New York: McGraw-Hill.

Eppelsheimer, D. 1938. "The Development of the Continuous Strip Mill," London: *Journal of the Iron and Steel Institute* II.

Fisher, D. A. 1951. *Steel Serves the Nation: The Fifty Year Story of United States Steel*. New York: US Steel.

Fisher, D. A. 1963. *The Epic of Steel*. New York: Harper and Row.

Ford, Bacon and Davis. *Reports on the operations of the United States Steel Corporation*. (200 volumes 1936–38) (See especially vols. 99, 109 and 200.) Unpublished.

Fraser, C. E., and Doriot, G. F. 1932. *Analyzing Our Industries*. New York: McGraw-Hill.

Goodale, S. L. 1931. *Chronology of Iron and Steel*. Cleveland: Penton Publishing Company.

Harbord, F. W., and Hall, J. W. 1918. (but written c. 1903) *The Metallurgy of Steel. II Mechanical Treatment*. London: Charles Griffin.

Hoerr, J. P. 1988. *And the Wolf Finally Came: The Decline of the American Steel Industry*. Pittsburgh: University of Pittsburgh Press.

Hogan, W. T. 1954. *The Development of American Heavy Industry in the Twentieth Century*. Fordham University.

Hogan, W. T. 1971. *Economic History of the Iron and Steel Industry in the United States*. 5 vols. Lexington, MA: DC Heath.

Hogan, W. T. 1991. *Global Steel in the 1990s*. Lexington, MA: D.C. Heath.

Horner, J. G. 1911. "Rolling-Mill." *Encyclopedia Britannica*.

Howe, H. M. 1911. "Iron and Steel." *Encyclopedia Britannica*.

Ingham, J. N. 1991. *Making Iron and Steel*. Columbus: Ohio State University Press.

Iron and Steel Engineers. 1993. *Directory of Iron and Steel Plants*. Pittsburgh: AISE.

Irvin, W. A. 1938. "Changing Trends in the Steel Industry." *AISI Yearbook*. New York.

Jeans, J. S. 1902. *American Industrial Conditions and Competition*. London: British Iron Trade Association.

Jones, E. 1922. *The Trust Problem in the United States*. New York: Macmillan.

Kaplan, A. D. H., Dirlam, J. B. and Lanzillotti, R. F. *Pricing in Big Business: A Case Approach*.

Kennedy, E. D. 1941. *The Automobile Industry*. Reynal and Hitchcock.

Keystone Metals Database. 2010. *Modern Strip Production Technologies*.

Knox, H. A. 1943. *Development of the American Tinplate Industry*. (Reprinted from *Iron Age*).

Lamoreaux, N. R. 1985. *The Great Merger Movement in American Business, 1895–1904*. Cambridge: Cambridge University Press.

Lawrence, E. S. 1930. *The Manufacture of Steel Sheets*. Cleveland: Penton.

Lesley, J. P. 1859. *The Iron Manufacturer's Guide to the Furnaces, Forges and Rolling Mills of the United States*. New York: American Iron Association.

Macon, W. W. 1922 and 1926. "Iron and Steel." *Encyclopedia Britannica*.

Malborn, J. 1950. "Hot Strip Rolling Mills: The Evolution of the Modern 4-High Mill." *Sheet Metal Industries* (UK).

Mathesius, W. 1952 "Iron and Steel Production and the Coke Industry." Paper to *Minerals Industries Symposium of the Centennial of Engineering*, Chicago 8 and 9 September.

May, E. C. 1945. *Principio to Wheeling 1715–1945: A Pageant of Iron and Steel*. New York: Harper&Brothers.

McLaughlin, G. E. 1938. *Growth of American Manufacturing Areas*. Pittsburgh: University of Pittsburgh Press.

Mitchell, W. C., and others. 1932. *Recent Social Trends in the United States*. New York: McGraw-Hill.

Morgan, C. H. 1901. "Some Landmarks in the History of the Rolling Mill" *TAIME* XXII 1901, pp. 31–64.

Nourse, E. G., and associates. 1934. *America's Capacity to Produce*. Washington: Brookings Institution.

Paskoff, P. (ed.). 1989. *The Iron and Steel Industry in the Nineteenth Century: Encyclopedia of American Business History and Biography*. New York: Facts on File.

Republic Steel Corporation. c. 1966. pamphlet "Central Alloy District."

Rodgers, A. L. 1949. *The Iron and Steel Industry of the Mahoning and Shenango Valleys*. Unpublished PhD, University of Wisconsin.

Rostow, W. W. 1960. *The Stages of Economic Growth*. Cambridge: Cambridge University Press.

Rostow. W. W. 1971. *Politics and the Stages of Growth*. Cambridge: Cambridge University Press.

Rowe, F. H. 1938. *History of the Iron and Steel Industry in Scioto County, Ohio*. Columbus: Ohio State Archeological and Historical Society.

Russell, G. A. V. 1938. *Report on Observations made during a tour of Representative Steel Works in the United States* (March–April). Sheffield, England: United Steel Companies.

Russell, G. A. V. 1941. *Notes on Visit to the United States and Canada February–April 1941*. Sheffield, England: United Steel Companies.

Ruttenberg, H. J., and S. 1940. "War and the Steel Ghost Towns." *Harper's Magazine*, January, pp 147–155.

Schroeder, G. G. 1953. *The Growth of Major Steel Companies 1900–1950*. Baltimore: Johns Hopkins.

Seeley, B. E. 1994. *Iron and Steel in the Twentieth Century: Encyclopedia of American Business History and Biography*. New York: Facts on File.

Shannon, R. W. 1930. *Sheet Steel and Tinplate*. New York: Chemical Catalogue Company.

Stocking, G. W. 1954. *Basing Point Pricing and Regional Development: A Case Study of the Iron and Steel Industry*. Chapel Hill: University of North Carolina Press.

Tarbell, I. M. 1925. *The Life of Elbert H. Gary: The Story of Steel*. New York: Appleton.

Taylor, M. 1938. "Ten Years of Steel." Statement at annual meeting of US Steel Stockholders, 4 April.

Temin, P. 1964. *Iron and Steel in Nineteenth Century America: An Economic Enquiry*. Cambridge, MA: MIT.

Tiffany, P. 1988. *The Decline of American Steel: How Management, Labor and Government Went Wrong*. New York: Oxford.

Tytus, J. B. *Technology and Culture*, 29 September 2012.

United States Steel Corporation. Notes on "Origin of Farrell Works" 1945; "History of "New Castle and Shenango works" n.d.; "History of Fairfield works" n.d.; "Brief History of "Vandergrift plant" n.d. unpublished; all in USS files.

Ward's Quarterly. 1967/1968. "Steel: A History" (a series of useful articles).

Warren, K. 1970. *The British Iron and Steel Sheet Industry since 1840*. London: Bell.

Warren, K. 1988. *The American Steel Industry 1850–1970*. Pittsburgh: University Press.

Warren, K. 2001. *Big Steel: The First Century of the United States Steel Corporation 1901–2001*. Pittsburgh: University Press.

Youngstown. 1950. *Fifty Years in Iron and Steel*. Youngstown: Youngstown Sheet and Tube Company.

Index

About the Author

Kenneth Warren died in June 2018. From 1979 he made numerous research visits to the United States, working on archives in Pittsburgh, Washington, and Bethlehem, Pennsylvania. In these visits he was given full access to the archives of Henry Clay Frick, and of the United States Steel Corporation and Bethlehem Steel Corporation. These researches resulted in books on United States Steel, Bethlehem Steel and the Connellsville coke district, and biographies of Henry Clay Frick and Charles Michael Schwab. After retirement from university teaching, he published twelve books, five of them on American subjects. The present study has drawn on a wide variety of sources, including industry-wide institutions and their publications, company records, and leading trade journals over more than a century.